U0162971

上海城市规划展示馆
Shanghai Urban Planning Exhibition Center

上海城市规划展示馆
参观指南

Guidebook of Shanghai Urban Planning
Exhibition Center

上海城市规划展示馆 编
By Shanghai Urban Planning
Exhibition Center

上海文化出版社

序言　上海：追求卓越的全球城市

习近平总书记指出："城市规划在城市发展中起着重要引领作用，考察一个城市首先看规划""规划建设管理都要坚持高起点、高标准、高水平，落实世界眼光、国际标准、中国特色、高点定位的要求"。

地处中国"江海之汇、南北之中"的上海，作为中国最大的经济中心城市、长江三角洲世界级城市群的核心城市、新兴的全球城市，在新时代将向卓越的全球城市、具有世界影响力的社会主义国际大都市的目标迈进。

国土空间规划是引领空间发展的指南，可持续发展的空间蓝图，是各类开发保护建设活动的基本依据。回顾上海城市发展进程，历版城市总体规划均对统筹协调社会经济、指导城市发展发挥了十分重要的作用。《上海市城市总体规划（2017—2035 年）》（简称"上海 2035"）是党的十九大后国务院批复的第一个超大型城市总体规划，也是新时代国土空间规划和治理改革的积极探索。规划着眼全球趋势、落实国家战略、立足上海实际，凝聚着广大市民对未来的美好憧憬和共同理想，是上海国土空间规划、建设和管理的基本依据和法定文件，是指导上海未来城市发展的纲领性文件，是实现"城市，让生活更美好"的发展蓝图。

立足新发展阶段、贯彻新发展理念、构建新发展格局，上海将始终铭记党中央赋予上海的新使命，认真践行"人民城市人民建，人民城市为人民"重要理念，在新的起点上继续发挥"改革开放排头兵、创新发展先行者"作用，聚焦"上海 2035"总体目标，持续推动城市高质量发展、创造高品质生活、实现高效能治理，不断满足人民群众对美好生活的向往，奋力创造新时代上海发展新奇迹。

Preface

Shanghai: Striving for the Excellent Global City

President Xi Jinping pointed out, "Urban planning plays an important leading role in urban development. When investigating a city, we must first look at its planning. We should adhere to a high starting point, high standards and a high level in both urban development and urban management, and implement the requirements of a global perspective, international standards, Chinese characteristics and high positioning."

Situated at the intersection of rivers and the sea, and the center of China's south-north route, Shanghai is China's largest economic center, the core city of the world-class Yangtze River Delta Urban Agglomeration and an emerging global city, striving to be an excellent global city and a socialist international metropolis with global influence in the new era.

Land space planning that provides a guide for spatial development and a blueprint for sustainable development is the basis for various development, conservation and construction. Reviewing the progress Shanghai has made in urban development, we can find that all Shanghai Master Plans released before have played a crucial role in coordinating the social and economic development and guiding urban development. Shanghai Master Plan 2017-2035 ("Shanghai 2035" for short) is the first megacity master plan approved by the State Council after the 19th CPC National Congress, and represents the active exploration of land space planning and governance reforms in the new era. Focusing on global trends and implementing national strategies based on Shanghai's real situation, while carrying out residents' visions and ideals for the future, the master plan is the underpinning and statutory document of land space planning, construction and governance, a programmatic document to guide

Shanghai's future development and a blueprint to turn the slogan "Better City, Better Life" into reality.

When implementing the new development concept and building a new development pattern based on the new development stage, Shanghai will always keep in mind the new mission assigned to the city by the CPC Central Committee, and earnestly implement the important concept that "The city is built by the people and for the people." It will continue to serve as a vanguard in reform and opening-up, and a pioneer in innovation-driven development at a new starting point, focusing on the overall goal of "Shanghai 2035" to promote high-quality development of the city, create a high-quality life, achieve high-efficiency governance, continuously satisfy people's yearning for a better life and strive to create a new development miracle in the new era.

上海城市规划展示馆入口
Entrance of Shanghai Urban Planning Exhibition Center

目录

Contents

11

3F An Innovative City

4F An Ecological City

Special Exhibits

Appendix

基本介绍

　　上海城市规划展示馆于 2000 年 2 月 25 日正式向公众开放，它是国内首家展示城市规划与城市建设成就的专业性展馆，围绕"城市、人、环境、发展"主题，全面呈现上海城市发展的昨天、今天和明天。

　　上海城市规划展示馆是全国 AAAA 级旅游景点、全国红色旅游景点、全国科普教育基地、上海市爱国主义教育基地、上海市对外文化交流基地、上海市志愿者服务基地，也是中共上海市委和上海市政府新闻发布会的承办单位。

　　上海城市规划展示馆通过二十余年的不断实践、探索和创新，被中外游客誉为"城市之窗""上海客厅"。

人民广场历史文化风貌区
People's Square Historical Cultural Area

About SUPEC

The Shanghai Urban Planning Exhibition Center ("SUPEC" for short) was officially opened to the public on February 25, 2000. It is the first facility in China that celebrates the achievements in urban revolution. Spinning around the people, environment and development of the city, it straddles its past, the present with the future.

The SUPEC is justly famous as an AAAA-level tourist destination, a revolutionary tourist attraction--each at the national level, and, a popular science education base, a patriotism education base, an international cultural exchange base, a volunteer service base--each at the municipal level. It is also a major venue for the press conferences of the CPC Shanghai Municipal Committee and Shanghai Municipal Government.

Over two decades of practice, exploration and innovation has earned the SUPEC the honors of the Window of Shanghai and the Living Room of Shanghai at home and abroad.

14

上海城市规划展示馆外景
Exterior of the SUPEC

更新改造

2017 年，国务院批复同意《上海市城市总体规划（2017—2035 年）》。为了全面展示上海新一轮城市总体规划所确定的城市发展目标和愿景，更好地激发社会共同行动，经上海市委、上海市政府同意，在上海市发展改革委员会、上海市财政局等相关部门的支持下，在上海市规划和自然资源局的指导下，上海城市规划展示馆更新改造工程于 2020 年 5 月全面启动。2022 年 8 月 9 日重新向公众开放。

本轮更新改造主题是"上海：追求卓越的全球城市"，即全面阐释上海努力践行习近平总书记提出的"人民城市人民建　人民城市为人民"的重要理念，充分展现上海建设卓越的全球城市和具有世界影响力的社会主义现代化国际大都市的远景目标，映射上海"海纳百川、追求卓越、开明睿智、大气谦和"的城市精神和"开放、创新、包容"的城市品格，充分表达人民对美好生活的向往和追求，以及上海全面落实国家战略的责任与担当。

本轮更新改造工程包括：展陈更新、建筑改造和智慧展馆建设。

建设中的上海城市规划展示馆
Shanghai Urban Planning Exhibition Center in construction

Renovation

With the consent of the Shanghai Municipal Committee and the Shanghai Municipal Government, SUPEC started the renovation project in May 2020, which was supported by relevant departments such as Shanghai Municipal Development and Reform Commission and Shanghai Municipal Finance Bureau and guided by Shanghai Municipal Bureau of Planning and Natural Resources, to fully demonstrate the goals and visions of urban development specified in the Shanghai Master Plan 2017–2035 ("Shanghai 2035"), which was approved by the State Council in 2017, and to drive the public to act together. The center was reopened to the public on August 9, 2022.

With the theme "Shanghai: Striving for the Excellent Global City," the renovation captures Shanghai's efforts to implement President Xi Jinping's proposal for "building a city by the people and for the people." The project seeks to reflect Shanghai's commitments to being an excellent global metropolis of socialism, relying on its inclusiveness, excellence, openness, modesty and innovation. It also expresses the people's desire for a better life, as well as the responsibility of Shanghai to carry out the national strategy.

The renovation scheme of the SUPEC aims to upgrade its displays, and the building while introducing a smart management system.

展陈更新

　　展陈更新全面展现"全球的上海、中国的上海、市民的上海"，深入浅出地解读上海立足新发展阶段、贯彻新发展理念、构建新发展格局，谋划建设更富魅力和更有温度的人文之城、更具活力和更加繁荣的创新之城、更具韧性和更可持续的生态之城，用世界语言讲述中国发展与上海创新先行的故事。

　　展览面积约 1.6 万平方米，分序厅、人文之城、创新之城和生态之城，包含 30 个展项，其中有近 500 张图片和图纸、45 个模型和近 100 部影像，主要应用的展陈技术：互动多媒体技术、全息展示技术、红外感应技术、AR/VR 技术，以及 5D 数字化沉浸式城市沙盘。

序厅中央环屏装置
Ring screen in the Preface gallery

Upgrading Displays

The upgraded displays seek to highlight the worldwide and nationwide status of Shanghai and what it means to its residents. They can tell you how Shanghai embrace new ideas and development approaches in the modern era, and what it undergoes to make it a charming human-centric culture-rich city, a dynamic, prosperous city of innovation, and a sustainable city of remarkable resilience. They are also a living memorial to China's development and Shanghai's creativity.

The exhibition space consists of four parts, namely, The Preface, A Humanistic City, An Innovative City, and An Ecological City, across approximately 16,000 square meters. The overall space is segmented into 30 exhibit areas, featuring some 500 pictures and drawings, 45 replicas and nearly 100 videos. Cutting-edge technologies are vying for your attention: interactive multimedia, holographic display, infrared sensors, AR/VR and 5D digital immersive sandtable.

二层历史文化风貌区模型
A 3D model of Historical Cultural Areas at the second-floor gallery

建筑改造

上海城市规划展示馆位于人民广场历史文化风貌区，本轮更新改造坚持"原址改造""建筑外形不变"的原则，以"简约适度、绿色低碳、健康智慧"为目标，力争使新建筑成为绿色和智能建筑的最佳实践案例。建筑改造主要包括功能布局完善、设施设备更新、绿色智能增设、景观设计优化等。改造后的建筑总面积为2万余平米，地下2层和地上6层。同时，对原地下一层的"1930风情街"进行了整体改建，改建后的主要功能是临展厅和城市规划公示区。

Improving Building

The SUPEC is ideally located in the People's Square Historical Cultural Areas. The renovation effort upholds the principle of "keeping the building at its existent site without changing its outlook"; while aiming to be a best example of green, smart building, the project refrained from a massive wave of engineering. The SUPEC now boasts better functional layout, equipment, and landscape design where green and smart devices are installed. The total area after renovation exceeds 20,000 square meters, with two basement floors and six above-ground. However, the former "Street circa. 1930" on the first basement floor (B1) has been remodeled into a temporary gallery and an area for urban planning announcement.

智慧展馆建设

　　智慧展馆系统平台是随着移动互联网技术发展而产生的大型交互平台，以移动互联网技术为基础，充分利用物联网技术、云计算、大数据分析等新一代信息技术，形成一个基于互联网的实时信息交互平台，实现展馆的智慧管理。

　　此次展陈力求为中外游客提供一个绿色、安全、高效、便捷和舒适的参观环境。在智慧互动领域，展馆提供了多样化的多媒体展示和交互技术，如多点触控、全息投影、体感互动、五维数字化沉浸式城市沙盘、三维投影等。在大数据管理及保护领域，建立了数据综合管理服务平台，主要包含展陈数据汇聚、分类管理、素材制作、展项管理与更新维护，用于建筑设施设备数据监测与分析管理的数十个智能控制子系统（主要包含能源、安防、消防、楼宇控制等）。在智慧服务领域，提供线上预约、线下智能导览、楼层导航、参观线路优选，以及参观信息分析和游客反馈意见受理等。

上海城市规划展示馆智能建筑系统界面
Interface of smart buildings of the SUPEC

Adding Smart Management

The smart management system is a large interactive platform that has emerged along with the progress of mobile Internet. This Internet-based real-time platform taps into the latest information technologies, such as Internet of Things, cloud computing, and big data analysis, to realize smart management of the SUPEC space.

The updated display expects to create a green, safe, efficient, convenient and comfortable environment for visitors from home and abroad. It draws on a dazzling array of multimedia display and interaction technologies, such as multi-touch, holographic projection, somatosensory interaction, 5D digital immersive sandtable, and 3D mapping. In addition, it features comprehensive data management, collecting, classifying and managing display data, producing materials, managing, updating and maintaining displays, with dozens of smart subsystems (mainly covering energy, security, fire protection and BA control) for data monitoring and analysis management of devices and equipment. There are also a mix of smart services, including online reservation, smart guides, floor navigation, recommended routes, as well as visitor information analysis and processing of visitor feedback.

转型提升

　　本轮更新改造主要在文案策划、空间设计、展项设计、展陈技术应用和策展组织方式五个方面进行了积极的探索和实践，主要的实施路径是"数据赋能"（可感知、能学习、善治理、会自造），"社群赋能"（建立共同愿景、培育共同爱好、形成群体智慧），以及"平台赋能"（学习和社交融合），提供参观者一个更开放、能交互、共创共赢的活动场所。

Transformation

The renovation focuses on five aspects, namely, copywriting, spatial design, design of exhibit areas, application of display technology and curation. It embraces the empowerment of the data (making information perceivable, enabling visitors to learn, govern and create), of the community (pooling collective wisdom based on shared vision and hobbies), and of the platform (combining learning with social activities). All these help make the SUPEC an open, interactive and win-win venue.

多媒体展示与交互技术
Multimedia display and interaction technology

文案策划

　　围绕更新改造的主题"上海：追求卓越的全球城市"，紧扣三个目标"人文之城""创新之城""生态之城"展开，每个目标依照"导则＋解读＋案例＋探索"的路径推进，帮助参观者更快、更准确地理解展览内容的逻辑关系和因果关系。同时通过对案例的进一步解读，激发参观者的兴趣，引导他们思考和实践探索。

Copywriting

To reiterate the theme "Shanghai: Striving for the Excellent Global City, " the three goals, namely, A Humanistic City, An Innovative City and An Ecological City, are unrolled with guidelines, interpretation, case study and exploration. Visitors can understand the logic underlying the exhibits and information. The interpretation text with cases can spark the interest of the visitors, and inspire them to think and explore.

互动学习装置
Interactive learning device in the exhibit area

空间设计

　　主要是从色彩、图形、装置入手，形成整个展馆的视觉体系，注重艺术氛围的营造，使整个参观环境更富艺术感和亲和力。比如：设计绿色和香槟金色作为展馆的代表色。"人文之城"选用橙色和圆形来呈现温度和融合，"创新之城"选用深灰色和三角形来体现科技感和未来感，"生态之城"选用绿色和方形来表达绿色低碳和生态网络。除此之外，把中国传统的"铜板雕刻＋喷砂＋烤漆＋手工做旧工艺"应用到入口的"城市大脑"艺术装置中，把上海的非物质文化遗产顾绣工艺植入到"城市肌理"展项，剪纸艺术融入到"人民城市"装置墙。色彩、图形和装置的有机结合，使参观者更容易识别和记忆，同时也提升了参观的趣味性。

Spatial Design

The visual system involves the color scheme, graphics and installations, hoping to create an artistic atmosphere in a welcoming environment. Green and champagne gold are the dominant colors of the SUPEC. Orange and circles are applied to highlight warmth and fusion in the section Humanistic City, dark gray and triangles to reflect the sense of technology and future in An Innovative City, and green and squares to indicate an eco-network and a low-carbon lifestyle in An Ecological City. In addition, techniques like the traditional Chinese copper plate carving, sandblasting, baking finish and aging are incorporated into the art installation "Urban Brain" at the entrance. The *Gu* embroidery, a type of intangible cultural heritage in Shanghai, is blended into the "Urban Fabric." The art of paper-cutting is found in the installation wall "People's City." The organic interplay of colors, graphics and installations makes it easier for visitors to identify and remember them, and also adds to the fun of the visit.

展项设计

　　紧紧围绕"教育＋体验＋探索",创新性地规划设计以"专业查询＋科普教育"为目的、"大咖＋小咖"(城市规划专业人士＋学生)共享互惠的"城市实验室"。它是全国首家基于真实地理信息数据的空间规划实验平台,开设了规划信息查询区"全球城市实践案例查询区""实验操作区""实验成果发布和分享区"。"城市实验室"的建立将全面提升展馆科普教育工作和规划公众参与活动的品质,同时能让参与者更有体验感、获得感和成就感。

Design of Exhibit Area

A shared "Urban Lab" for urban planning professionals and students alike is designed for the purpose of professional inquiry and popular science education based on a replica that combines education, experience and exploration. It is the first spatial planning experimental platform based on real geographic data in China. It consists of Inquiry about Planning of Shanghai, Global Information on Best Urban Practice Cases, Experiments, and Release of Experimental Results. The Urban Lab will comprehensively improve the quality of popular science education and planned public activities in Shanghai, while giving participants a better experience as well as a sense of achievement.

展陈技术应用

　　"人文之城"通过大量精致的微雕建筑模型来呈现历史和温度。"创新之城"应用三维投影、全息影像、五维数字化沉浸式城市沙盘等新技术来展现数字城市和未来城市。"生态之城"应用增强现实技术与虚拟现实技术等来营造绿色低碳、节能环保、生态宜居的可持续发展之城。

Display Technology

A large number of exquisite micro-carved architectural replicas are applied to highlight history and humanity in the section "Humanistic City." New technologies like 3D mapping, holographic projection, and 5D digital immersive sandtable in "Innovative City" present a digital city in the future. "An Ecological City" features augmented reality and virtual reality (AR/VR) to create a sustainable city that is low-carbon, energy-saving and livable.

多媒体展示与交互技术
Multimedia display and interaction technology

策展组织方式

　　本轮更新改造工程在上海市规划和自然资源局的领导下组织实施，展陈内容由局相关处室和专家团队负责审核。市区各级管理部门、各领域的专家学者、专业机构、高等院校，以及热心的市民都踊跃参与其中，使得展陈内容丰富、翔实、准确、精彩。

Curation

The renovation was organized and implemented under the leadership of the Shanghai Municipal Bureau of Planning and Natural Resources, and the displays are approved by related departments and expert teams. Urban management departments at all levels, experts and scholars from various fields, Professional institutes, colleges and universities, and enthusiastic citizens all actively participate in it, together creating a rich, accurate and exciting exhibition.

一层 **上海：追求卓越的全球城市**

1F　Shanghai: Striving for the Excellent Global City

一层展厅平面图
Plan of the first-floor gallery

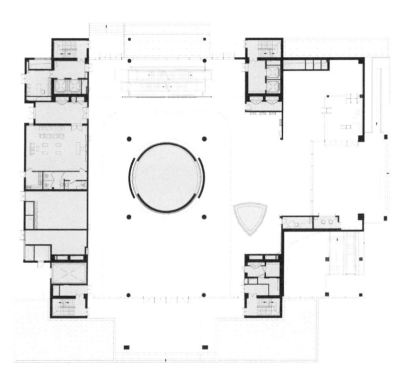

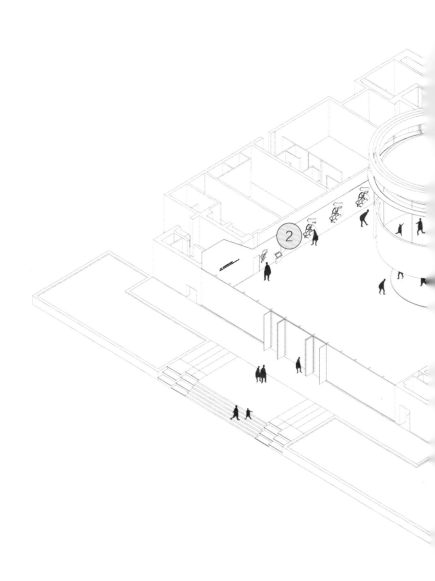

1 上海印象 Shanghai Impression
2 上海历次城市总体规划概述 Previous Versions of Shanghai Master Plan
3 人民城市艺术装置 Art Installations under the Theme of the City
4 城市大脑 City Brain
5.1 智慧展馆展示 Smart Exhibition System
5.2 接待和智慧服务站 Reception and Intelligent Service Center

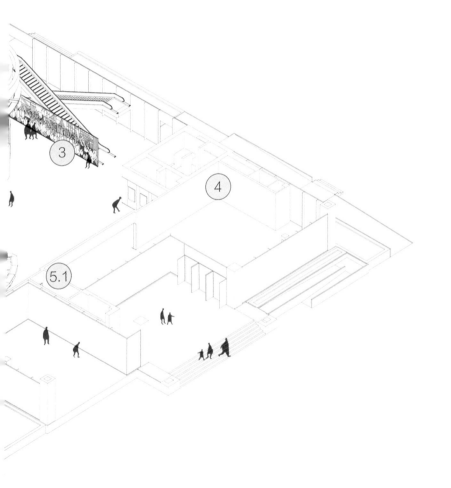

地处中国"江海之汇、南北之中"的上海，作为中国最大的经济中心城市、长江三角洲世界级城市群的核心城市、新兴的全球城市，在新时代将向卓越的全球城市、具有世界影响力的社会主义现代化国际大都市的目标迈进。序厅以"上海：追求卓越的全球城市"为主题，通过"上海印象""上海历次城市总体规划历程概述""人民城市艺术装置""三个展项，多维度呈现上海立足新发展阶段、贯彻新发展理念、构建新发展格局，聚焦"上海2035"总体目标，持续推动城市高质量发展、创造高品质生活、保障高效能运行，不断满足人民群众对美好生活的向往，奋力创造新时代上海发展新奇迹。

序厅作为参观流线的起点，以圆形的中央环屏装置为中心，观展空间环绕中心布局。"上海印象"将上海的发展历程和"上海2035"的核心愿景以电影艺术融合数字技术的方式创新性演绎，拉开整体展陈的序幕。"上海历次城市总体规划概述"与"人民城市艺术装置"展项分别在西、北两侧展开，前者解读上海总规历程和发展蓝图，后者用高度凝练与抽象的艺术语言回应"人民城市"的重要理念。展厅东侧设有接待中心和智慧服务站，为观众提供咨询、导览等服务。

序厅以纯净的白色为主基调，选用玻璃、铝板等环保材料，力图在纯粹的展览环境中烘托出主题鲜明且多元的展示内容，营造纯净、灵动、智慧、亲切的城市体验空间。

Shanghai, located halfway along China's eastern coastline, is the largest economic hub in China, the core of the world-class urban cluster in the Yangtze River Delta, and an emerging global city. It is striving to be an excellent global city and a socialist modern international metropolis with world influence in the new era. With the theme "Shanghai: Striving for an Excellent Global City," The Preface features three exhibit areas, namely, a dynamic interpretation of "Shanghai Impression," a 3D presentation of "Previous Versions of Shanghai Master Plan" and a core installation depicting "People's City." In the new era, Shanghai will adopt new ideas of and approaches to development in response to the "Shanghai 2035." It will continue to drive development, create quality life, and ensure efficient operation, constantly satisfying the people's yearning for a better life, and striving for new breakthroughs in the new era.

In the Preface, the tour starts with the ring screen at the center. The exhibit area of "Shanghai Impression" innovatively chronicles the history of Shanghai and the outlook of "Shanghai 2035." It integrates digital technology into cinematic art, making for a fitting curtain-raiser to the whole exhibition. On the west and the north are the exhibit areas of "Previous Versions of Shanghai Master Plan" and "Art Installations under the Theme of the City." The former consists of the master plans and development blueprints of Shanghai over the years, while the latter interprets the core idea, "A People's City," in a highly condensed and abstract artistic language. The reception and smart service station in the east can answer the questions of visitors and arrange guided tours.

Pure white being the dominant color, the Preface adopts green materials with smooth surfaces, such as glass and aluminum panels to correspond to the theme and the diverse holdings; it creates a pristine, fluid, smart, and cordial ambience.

上海印象 Shanghai Impression

星空穹顶实景图
The starry sky dome

展示内容

　　"上海印象"为全馆展陈的序章。展项将《上海：追求卓越的全球城市》《时空之轮》和《梦想之城》三部影片通过多种数字演绎模式，多维度呈现上海的过去、现在与未来。

　　《上海：追求卓越的全球城市》主题影片，以历版城市总体规划为脉络，讲述上海自开埠以来城市发展和空间格局演变的历程，描摹着"上海2035"，展现出卓越全球城市的美好未来。从中可以看到规划对上海城市空间格局的引领和推动，以及一代代人在都

市空间中奋力逐梦，与这座国际大都会共同成长奋进的图景。

《时空之轮》围绕上海外滩的变迁，通过叠屏模式，以六幕时间切片跨越上海开埠至今的历史变迁，承载起穿越百年的城市记忆。

《梦想之城》围绕"上海2035"确定的生态之城、人文之城、创新之城三个主要目标，以环幕模式展现生态、人文、创新的一幕幕动态长卷，以写意动画凸显城市气质，勾勒美好蓝图。

Content

"Shanghai Impression" preludes the entire exhibition. Mingling various digital modes, the three-part film consists of "Shanghai: Striving for the Excellent Global City," "Time and Space Wheel" and "Dream City," spanning the past, present and future of Shanghai.

"Shanghai: Striving for the Excellent Global City", based on the master plans of Shanghai, presents its urban and spatial development since it was opened as a Treaty Port in 1843, as well as the rosy future of Shanghai as an excellent global city as depicted in "Shanghai 2035." It unravels how the master plans propel the development of urban space, where generations after generations strive for their dreams and prosper alongside the international metropolis.
The film "Time and Space Wheel" traces the evolution of the Bund since the opening-up of Shanghai in1843 in six scenes, serving as a memory of the city over a hundred years.

The film "Dream City" is centered on the goal of turning Shanghai into an ecological city, culture and innovation, as specified in "Shanghai 2035." The dynamic footage is unraveled on the ring screen, showcasing the aura of the city and a promising blueprint with animation of freehand drawings.

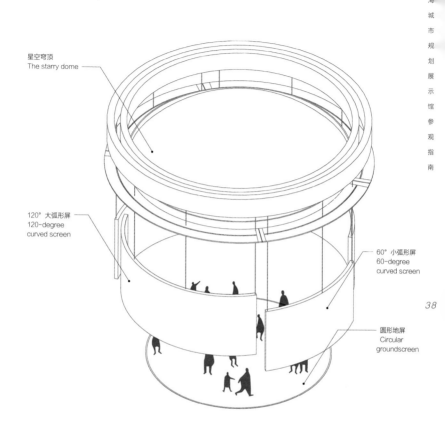

星空穹顶
The starry dome

120°大弧形屏
120-degree
curved screen

60°小弧形屏
60-degree
curved screen

圆形地屏
Circular
groundscreen

展项解读

　　"上海印象"设有大型双面弧形 LED 环屏装置，环绕中央穹顶悬置，纵向贯穿首层与夹层中庭空间。

　　装置由处于内环的两块 120°大弧形屏、处于外环的两块 60°小弧形屏及一块圆形地屏构成，环绕式结构为观众提供 360°全方位观看视角。组合屏装置所产生的空间包裹感和影像沉浸感，使观众身临其境地感受城市愿景。

星空穹顶的设计融合悬挂式环屏，自然连接屏下空间与周边展陈空间，巧妙地将夹层空间融为一体，提升了建筑空间的指向性与通透性。

展项设置了单屏、环屏、叠屏及地屏等多种数字演绎模式。单屏模式通过一块 120°大弧形屏和两块 60°小弧形屏互动，形成一个动态的环幕影院，生动展映《上海：追求卓越的全球城市》。叠屏模式通过四块弧形屏高低错落的动态运转及地屏的时空罗盘演绎，构成立体震撼的"时空视窗"，形象展映《时空之轮》。环屏模式即通过四块弧形屏完整闭合，形成一个全角度环绕的立体环幕空间，动态演绎《梦想之城》。地屏模式将四块弧形屏全部向上收归至顶部，通过约 12 米直径的圆形地屏完整演绎上海的空间地理影像。

各种模式能够根据具体展示需求进行快速切换并有效呈现，为数字化更新展示创造更多可能。

"上海印象"采用大型 LED 环屏装置
LED ring screens in the exhibit area "Shanghai Impression"

Interpretation

This area is equipped with large double-sided curved LED ring screens, hung around the dome and run vertically through the first floor and the mezzanine atrium.

There are two 120-degree curved screens in the inner ring, two 60-degree curved screens in the outer and a circular floor screen. Such a combination provides visitors with an all-round view. When they are wrapped inside the screens, they will gain an immersive experience of the city's visions.

This area hosts a variety of digital modes on its single, ring, multiple screens and one on the ground. For the first mode, a large 120-degree curved screen and two small 60-degree ones interact to form a dynamic theater that vividly displays the film "Shanghai: Striving for the Excellent Global City." The second film "Time and Space Wheel" is presented with multiple screens, that is, four curved screens rotating at different heights, alongside the interpretation over space and time on the ground screen for a striking 3D view. The ring screen consists of four completely connected curved screens forming a three-dimensional ring to dynamically interpret "Dream City." When the ground screen mode is on, all the four curved screens are brought up to the top; a full map of Shanghai is revealed on a ground screen, which is a circular fixture in a diameter of about 12 meters.

The modes can be quickly switched and work effectively as needed, allowing it possible to digitally update the display.

各模式组合结构图、屏幕运动分析图 Diagram of four rotation modes

单屏模式
Single screen

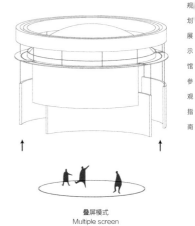

叠屏模式
Multiple screen

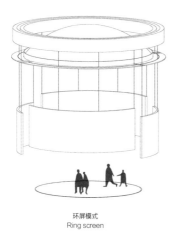

环屏模式
Ring screen

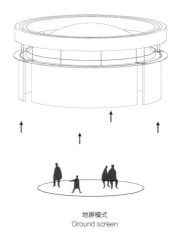

地屏模式
Ground screen

技术特色

"上海印象"展项采用大型弧形屏多模式组合的沉浸式展示技术,属全国首创。单块 120° 弧形屏弧长 12 米、高度 3.3 米、重约 4 吨,单块 60° 弧形屏弧长 6 米、高度 3.3 米、重约 2.5 吨,四块弧形屏总重近 13 吨。为避免悬挂的厚重弧形屏幕在顶部空间有限的室内环境中形成空间压抑感,设计将超大面积的双面显示屏的总厚度由 26 厘米优化至 18 厘米,减薄了 31%,呈现出轻盈的视觉效果,而且每一个单元均设置防脱落保护装置,有效保障弧形屏安全稳定运行。

屏幕环形运动的实际运行速度能够达到 0.5 米/秒。为缓解弧形屏体高速运动时产生的惯性冲击力,设计采取软链接技术,基于定制的环形运动轨道上,有效实现升降 6 米高度且能稳定运转的展示需求。结构通过控制钢丝绳拉伸角度和电机算法设计,解决了高速环形运动中的晃动问题,使屏体在升降过程中稳定运转,避免倾斜的现象。

四块弧形屏不仅需要超 360° 旋转,同时需要实现 6 米的上下升降,在立体空间里做三维运动,这对于藏于穹顶上方的机械结构是极大的挑战。为应对这一难题,设计研发了全套环形机械运行系统、升降系统和滑轮组,在同步顺滑升降和旋转的同时,实现弧形屏低噪声运动、保障观赏体验,并成功将噪声控制在 55 分贝以下,仅相当于台式电脑主机运行的声音。此外,设计在屏幕的升降、旋转轨道加装条形读码、光电、感应、计数器等多种与自动保护装置相联动的传感器,实时监控,及时反馈调整。在实现材料高承载、轻晃动、稳运转、低噪声等要求的同时,通过"绝对闭环控制"搭建逻辑,有效补偿技术误差。

Technical Features

The exhibit area "Shanghai Impression" is equipped with a multi-mode combination of large curved screens—the first one in China—that adopts the immersive display technology. A single 120-degree curved screen is 12 meters long and 3.3 meters high, and weighs about 4 tons. A single 60-degree curved screen is 6 meters long and 3.3 meters high, and weighs about 2.5 tons. The total weight of the four curved screens nears 13 tons. To avoid a depressing sense created by the heavy suspended curved screens in the limited indoor space, the total thickness of the super-large double-sided screen is reduced from 26 cm to 18 cm, down by 31%, for a sense of lightness. Moreover, protection devices are in place to effectively ensure safe and stable operation of the curved screens.

The actual speed of the screen during circular movement can reach 0.5m/s. To alleviate the inertial impact generated by its high-speed motion, soft link technique is adopted based on the customized circular track to effectively ensure that a screen runs stably while being lifted up and down by 6 meters. The shaking problem in the high-speed circular motion is addressed by controlling the stretching angle of the wire rope and designing motor algorithms, so the screen can run stably during the lifting process without tilting.

The four curved screens not only need to rotate over 360 degrees, but also need to move up and down by 6 meters, and perform three-dimensional movement in a three-dimensional space, which is a daunting challenge for the mechanical structure hidden above the dome. To cope with it, a complete set of circular mechanical operation system, lifting system and pulley block is designed and developed. During lifting and rotation, the noise of movement is reduced to under 55 decibels, which is equivalent to the sound of the mainframe of a desktop computer, and a good viewing experience is guaranteed. In addition, a variety of sensors linked with automatic protection devices, such as bar code reader, photoelectric sensor, inductor and counter, are integrated into the lifting and rotating tracks for real-time monitoring as well as prompt feedback and adjustment. While meeting the requirements of a high load, less shaking, stable operation, and low noise, the logic is built through "absolute closed-loop control" to effectively offset technical errors.

上海历次城市总体规划概述
Previous Versions of Shanghai Master Plan

"上海历次城市总体规划概述"轴测图

Axonometric drawing of "Previous Versions
of Shanghai Master Plan"

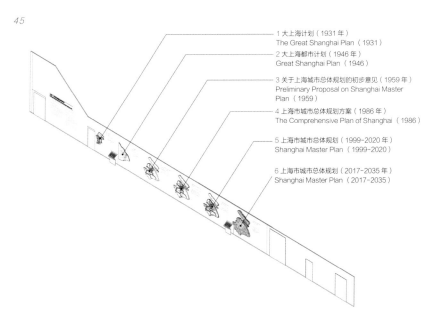

1 大上海计划（1931 年）
The Great Shanghai Plan （1931）

2 大上海都市计划（1946 年）
Great Shanghai Plan （1946）

3 关于上海城市总体规划的初步意见（1959 年）
Preliminary Proposal on Shanghai Master
Plan （1959）

4 上海市城市总体规划方案（1986 年）
The Comprehensive Plan of Shanghai （1986）

5 上海市城市总体规划（1999-2020 年）
Shanghai Master Plan （1999-2020）

6 上海市城市总体规划（2017-2035 年）
Shanghai Master Plan （2017-2035）

展示内容

　　"上海历次城市总体规划概述"以时间为脉络，以城市规划为主线，集中展示上海自 1931 年"大上海计划"起至今的六轮城市总体规划，重点突出"上海 2035"这一凝聚着广大市民美好憧憬和共同理想的新一轮发展蓝图，生动展示上海在新时代向卓越的全球城市、具有世界影响力的社会主义现代化国际大都市的目标迈进。

　　国土空间规划是引领空间发展的指南，可持续发展的空间蓝图，是各类开发保护建设活动的基本依据。在此通过回顾上海城市发展进程，讲述历版城市总体规划对统筹协调社会经济、指导城市发展中所发挥的重要作用，体现新形势下上海在国土空间规划和治理改革中的积极探索。

Content

This area presents the six successive master plans of Shanghai, spanning from the "The Great Shanghai Plan" dating to 1931 to "Shanghai 2035," the most recent blueprint that carries the public's vision and yearning for a better future. It demonstrates Shanghai's goal of becoming an excellent global city and a socialist modern international metropolis with world influence in the new era.

Land planning is a guide for spatial development, a spatial blueprint for sustainable development, and the basis for various types of development, protection, and construction activities. A review of the development of Shanghai reveals the important role of previous urban master plans in coordinating economic activities and guiding urban development, reflecting Shanghai's active exploration in the reform of land planning and governance under the new situation.

大上海计划（1931 年） 1927 年上海特别市成立后，上海特别市政府即着手编制上海城市历史上第一个综合性都市发展总体规划。规划突出上海作为港口城市特点，在空间布局上跳出当时已建成的租界和华界地区，选择在江湾一带规划新的市中心。

The Great Shanghai Plan (1931). After the Shanghai Special City was established in 1927, the Shanghai Special City Government began to prepare the first comprehensive master plan in the history of Shanghai. The plan highlights the characteristics of Shanghai as a port city and locates the new civic center in Jiangwan instead of the then established foreign settlements or other area.

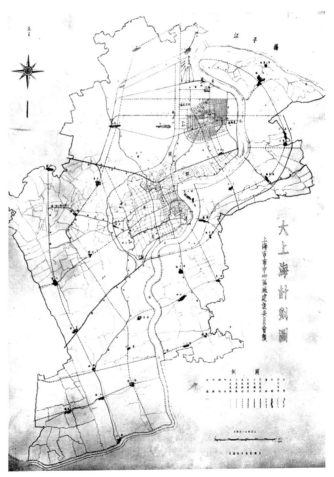

大上海计划总图初稿
Map of the Great Shanghai Plan

大上海都市计划（1946 年） 1945 年抗日战争胜利后，于 1946—1949 年间编制三稿大上海都市计划，规划着眼未来五十年发展，提出"港埠都市，亦将为全国最大工商业中心之一"定位。在空间布局上，以区域主义视角将规划范围拓展到上海特别市周边的江苏、浙江区域，同时运用"田园城市"规划理念，提出"有机疏散、组团结构"的布局思路。

Great Shanghai Plan (1946). After the victory of the War of Resistance Against Japanese Aggression in 1945, three drafts of the Shanghai Master Plan were drawn up between 1946 and 1949, focused on the development over the next five decades. It positioned the port city as one of the largest industrial and commercial centers in China. Broader regional planning, with the area expanding to Jiangsu and Zhejiang area at the time, was combined with the idea of a "garden city" that features organic evacuation and a cluster structure.

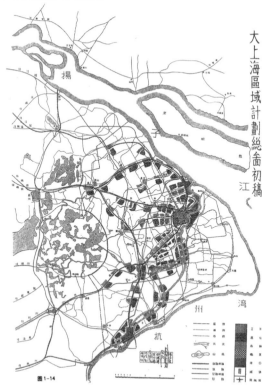

大上海区域计划总图初稿
Planning Draft of the Great Shanghai Area

关于上海城市总体规划的初步意见（1959 年） 1958
年国家分两批将江苏省十个县纳入上海行政辖区范围
后，于 1959 年编制此版规划。为支撑上海作为重工
业城市的定位，在空间布局上一方面逐步改造旧市区，
严格控制近郊工业区以避免城市蔓延；另一方面在郊
区重点规划工业卫星城。

Preliminary Proposal on Shanghai Master Plan (1959).
This version was formulated in 1959 after ten counties
in Jiangsu were included in two batches by the country
in the jurisdiction of Shanghai in 1958. In response to
the positioning of Shanghai as a powerhouse of heavy
industry, the spatial layout of the old urban area was
gradually transformed to strictly control the suburban
industrial area and prevent urban sprawl, while the planning
of industrial satellites in the suburbs was highlighted.

上海区域规划示意草图（1959）
Sketch of Shanghai's Spatial Planning

上海市城市总体规划方案（1986 年） 作为改革开放以后国务院对上海批复的第一个法定性城市总体规划，再次强调上海作为太平洋西岸最大的经济和贸易中心之一的开放门户定位。在空间布局上延续以往历版总体规划明确的中心城改造和卫星城建设并举的方略，以及充实发展卫星城的思路，同时首次将南北两翼发展和浦东地区开发纳入规划蓝图中。

The Comprehensive Plan of Shanghai (1986). As the first legal master plan approved by The State Council after the reform and opening up, it once again emphasized the positioning of Shanghai as one of the largest economic and trade centers on the west coast of the Pacific Ocean. While the strategy of reconstructing the central city and promoting the satellite towns in previous plans was continued, the development of the north and south wings as well as the Pudong area were included for the first time.

上海市城市总体规划图
Map of Shanghai Master Plan

上海市城市总体规划（1999—2020 年） 为适应 1990 年代浦东开发开放和洋山深水港建设的需要，这版总体规划明确上海国际大都市的定位和国际经济、金融、贸易、航运中心的发展目标。在空间布局上明确了中心城—新城—重点镇——一般镇的城镇体系，拓展了城市沿江、沿海发展的空间。

Shanghai Master Plan (1999-2020). To meet the needs of the development and opening of Pudong and the construction of Yangshan Deep Water Port in the 1990s, this Master Plan defined Shanghai as an international metropolis and an international hub of economy, finance, trade and shipping. It specified the urban system of the central city, new cities, central towns and market towns, and expanded the room for development along the rivers and the coast.

上海市城市总体规划图（土地使用规划）
Map of Shanghai Master Plan (Land Use Plan)

上海市城市总体规划（2017—2035 年） 围绕国家既定的"两步走"方略，展望 2035 年的城市蓝图，这版总体规划提出卓越全球城市的目标远景和创新之城、人文之城、生态之城三个子目标以及对应规划策略。在空间布局上则强调立足区域视角，形成"网络化、多中心、组团式、集约型"的空间体系。

Shanghai Master Plan (2017-2035). In response to China's two-step approach and the prospect of the year 2035, this Master Plan puts forward the vision of Shanghai as an excellent global city, with three sub-goals, namely, an innovative city, a humanistic city, and an ecological city, as well as corresponding strategies. From a regional perspective, it specifies a "network-based, multi-center, clustered, and intensive" spatial system.

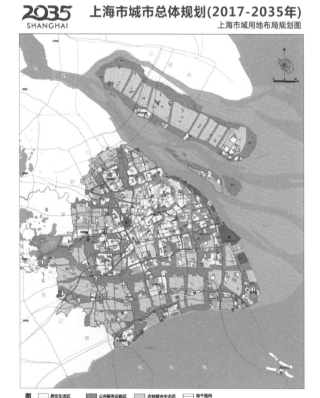

2035 SHANGHAI
上海市城市总体规划(2017-2035年)
上海市域用地布局规划图

上海市域用地布局规划图
Shanghai Municipality Spatial Layout Map

展项解读

　　"上海历次城市总体规划概述"中，六张总体规划图以白色人造石为基底，采用曲线流畅的分层亚克力板塑形，图面采用分层浮雕工艺制作，镌刻出丰富且厚重的时间叠层。同时画面肌理与色彩从1931年"大上海计划"的复古木色纹理逐步过渡至"上海2035"彩色图版，在立体还原总规图纸细节的同时，凸显历次"总规"的发展变化。

　　两侧的立式查询屏，是对更翔实的图纸与历史影像的延展，为公众深度搭建起连接城市过去、现在与未来的桥梁。

Interpretation

In this area, the six master plans are created on layered acrylic panels with smooth curves with a white artificial stone base. Scenes in layered relief indicate the rich evolution over time. The texture and color gradually transit from the retro wood-colored texture for the "Great Shanghai Plan" in 1931 to the colored plate for "Shanghai 2035." The restoration of the details of the master plan drawings in three dimensions also highlights the development and changes over time.

The vertical query screens on both sides offer more detailed drawings and historical images, connecting the past, the present and the future of the city for the public.

人民城市艺术装置
Art Installation under the Theme of the City

"人民城市艺术装置"轴测图

Axonometric drawing of "Art Installation under the Theme of the City"

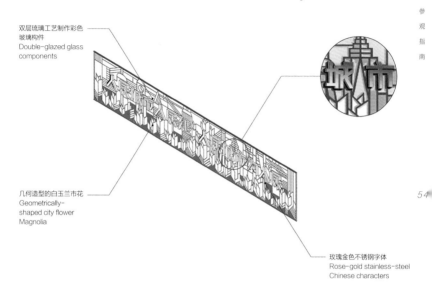

双层琉璃工艺制作彩色玻璃构件
Double-glazed glass components

几何造型的白玉兰市花
Geometrically-shaped city flower Magnolia

玫瑰金色不锈钢字体
Rose-gold stainless-steel Chinese characters

54

人民城市艺术装置设计稿
Drawing of Art Installations under the Theme of the City

展示内容

　　"人民城市人民建，人民城市为人民"是习近平总书记在 2019 年考察上海时提出的重要理念。"人民城市艺术装置"展项以"人民城市"的理念为核心，通过极具海派风格的剪纸艺术，将几何造型的白玉兰市花，错落有致地与城市景观和文化、自然等要素进行叠合，在突显上海多元特色与独特魅力的同时，呈现出一幅充满活力与温度的社会主义现代化国际大都市的景象。

Content

"The city is built by the people and for the people." This is an important concept raised by President Xi Jinping when he inspected Shanghai in 2019. Based on it, the installation well blends the Shanghai-style paper-cutting with the geometrically shaped city flower Magnolia, urban culture, innovation, nature and other elements. While accentuating the diversity and unique charm of Shanghai, it depicts a socialist modern international metropolis full of vitality and warmth.

展项解读

　　"人民城市艺术装置"位于中庭中央轴线上，融合了海派剪纸、版雕和油画的艺术手法，将传统与现代的美学张力融汇在"人民城市"的重要理念之中。

　　匠心传递初心，设计者海派重彩画家李守白先生从城市建筑中提炼出六种色彩元素，如摩登复古的墨绿色、石库门里弄的砖红色、温暖石材的黄色等，让人文情怀融入多彩都市。为更精细地呈现装置色彩，展项运用双层琉璃工艺制作彩色玻璃构件，并依照艺术家图稿切割为面积相异的几何形态，嵌入轻钢龙骨骨架中。玫瑰金色的不锈钢字体——"人民城市人民建 人民城市为人民"悬挂于晶莹剔透、错落有致的城市背景之上，徐徐展开城市的美丽画卷，展现着独属于这座城市的人文魅力。

Interpretation

The installation is located on the central axis of the atrium. It combines the artistic techniques of Shanghai-style paper-cutting, engraving and oil painting, and integrates the interplay between traditional and modern aesthetics into the important concept "People's City."

The original aspiration is conveyed with ingenuity. The designer, Mr. Li Shoubai, a Shanghai-style heavy-color painter, derives six colors from buildings in Shanghai, including the modern and retro dark green, the brick red in Shikumen Lane, and the warm stone yellow, which adds a touch of nostalgia into the colorful city. For a more delicate presentation of the colors, the exhibit area features the double-glazed glass components, which are cut into geometric shapes of varied area according to the artist's drawings and embedded in the light steel frame. The rose gold stainless steel for the sentence "the city is built by the people and for the people" in Chinese characters hangs against the crystal clear and well-proportioned background, slowly unfolding the beautiful scroll of the city to show its unique charm.

城市大脑　City Brain

展示内容

　　未来的智慧化城市是类似于人类大脑神经元网络的多中心化结构，节点之间联系度与节点自身活跃度的提升将大大提高城市网络的活力和智慧度。展项位于序厅东侧，由上海的"城市大脑"衍生而来，展示了未来数字化城市的多中心结构，如同人类大脑的神经元网络，隐喻上海以"绣花针"般精密的算法支撑起精细化的城市治理。

Content

Located in the eastern section of the Preface, this inhibit area is inspired by the "City Brain" project in Shanghai. Smart cities in the future feature a multi-center structure similar to the neuron network of human brain. The connection between the "neurons" and their activity decide the vitality and intelligence of these cities. The special structure indicates that the elaborate urban governance in Shanghai is supported by rigorous and precise algorithms.

城市大脑
"City Brain" exhibit area

展项解读

"城市大脑"由两面高 2.8 米，分别长 10.6 米与 7.35 米，呈 L 形布局的艺术墙面组成。创作采用复杂的铜版雕刻工艺，历时数月时间，将千丝万缕、粗细不一的铜线以及节点，精致地打磨出来。铜线中藏有黄浦江、苏州河的蜿蜒形态。绿色为底，金色为线。设计采用金与绿两种铜的不同状态，它们既是港口城市特有的代表色，同时又是中国文化的典型装饰元素。两者的结合运用，展现出上海中西合璧、传统与现代并存的人文气质。

Interpretation

The L-shaped installation consists of two 2.8-meter-tall art walls, with a length of 10.6 meters and 7.35 meters, respectively. Drypoint is applied to create the golden meandering Huangpu River and Suzhou Creek against a green backdrop with copper wires of different sizes. The two colors are unique to port cities and typical decorative elements in China. Together they demonstrate the marriage between Chinese and Western cultures as well as between the tradition and the modern in Shanghai.

城市大脑细节
Details of City Brain

二层展厅平面图
Plan of the second-floor gallery

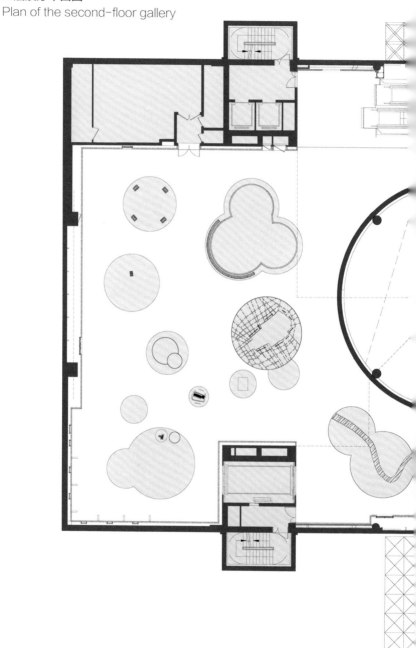

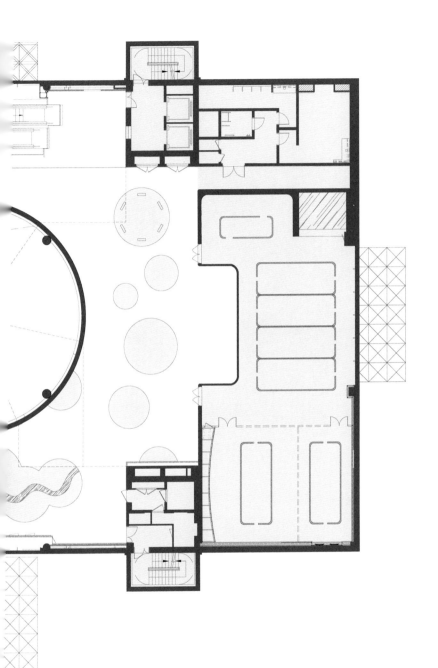

1 城市肌理 Urban Fabric
2 风貌格局 Feature Patterns
3 城市更新 Urban Regeneration
4 黄浦江、苏州河——迈向具有全球影响力的世界级滨水区
The Huangpu River and Suzhou Creek—World-class Waterfronts
5 城市设计标准 Urban Design Standards
6 15 分钟社区生活圈
15-minute Community Life Circle
7 公共服务 Public Service
8 公众参与 Public Participation

上海是国家历史文化名城，中国最具影响力的国际大都会之一。江南文化是上海文化的底色，海派文化是上海文化的特色，红色文化是上海文化的亮色。二层展厅以"人文之城"为主题，突出"以人为本"的城市发展理念，重点凸显红色文化、海派文化、江南文化。

文化，是一座城市的气质、风骨和灵魂。上海这座城市将"人"视为最宝贵的发展资源，是历史文化的积淀所在，更是一切发展的价值旨归。"人文之城"通过城市肌理、风貌格局、城市更新、黄浦江与苏州河——迈向具有全球影响力的世界级滨水区、城市设计标准、15分钟社区生活圈、公共服务、公众参与八个展项，从上海悠久的城市文脉、建立历史文化风貌保护体系、重塑高品质公共空间、以生活圈为单元的社区治理、完善和保障多层次公共服务，再到人们对上海这座城市的归属感和认同感，全方位、多角度呈现上海对于提升城市品质的不懈坚守，彰显独属于上海的"海纳百川、追求卓越、开明睿智、大气谦和"的城市精神，以及建设更具人文底蕴和时尚魅力的国际文化大都市的美好追求。

二层展厅各展项围绕中央环形空间布置，呈散点状排布。展厅的视觉设计选用纯粹且完整的圆形作为基础元素，并扩展出弧形、环形、圆形等同源形态并互相组合。通过对展台、地胶、灯具等形态大小的变化或形体交叠，自然分隔出不同展区空间。展厅应用木质、丝质、复合型人造石等有机材料，在桔灰色基调的展陈环境中营造温暖、轻盈、柔和的空间氛围。

Shanghai is famed as a culture-rich city and a major international metropolis in China. It is a remarkable tapestry of the culture in the Yangtze Delta and a depository of wealth that recalls the revolutionary past of China. Themed "A Humanistic City," the second-floor gallery focuses on Shanghai's human-centric urban development, paying tribute to its revolutionary history and its hybrid culture.

Culture is the underlying spirit of a city. Shanghai sees "people" as its most valuable resource for development. It is people that have shaped the history and culture of the city, and no discussion about development can make sense without a relevance to its people. The eight exhibit areas of the gallery are dedicated to Shanghai's urban fabric, planning, renewal, waterfront along the Huangpu River and the Suzhou Creek, urban design standards, 15-minute community life circle, public services, and public participation. They inspire the visitors to learn more about Shanghai's social history, how it establisheds its heritage conservation system and high-quality public space, how it manages its living circle-based community governance, improves and multi-level public services, and where people's sense of belonging and identity come from. The holdings testify to Shanghai's commitment to improving the quality of the city, and celebrate its all-inclusive spirit, its ambition to aim high, and its readiness to look out to the rest of the world.

The holdings on the second floor are clustered around the circular wall at the center. The ubiquitous circles prove to be the dominant visual elements, and derive into a slew of homologous forms: arcs, rings and circles that connect with each other. Round display tables, flooring, lamps in different diameters overlap, ingeniously separating the entire space into different exhibit areas. Organic materials—wood, silk and composite artificial stone—evoke a cozy, light and soft touch in an orange-gray setting.

城市肌理　Urban Fabric

"城市肌理"展项轴测图
Axonometric drawing of "Urban Fabric"

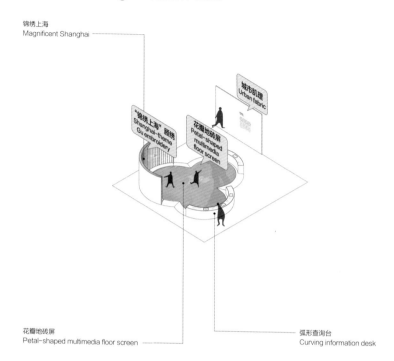

锦绣上海
Magnificent Shanghai

"锦绣上海"顾绣
Shanghai-theme
Gu embroidery

花瓣地砖屏
Petal-shaped
multimedia
floor screen

城市肌理
Urban fabric

花瓣地砖屏
Petal-shaped multimedia floor screen

弧形查询台
Curving information desk

展示内容

在漫长的历史进程中，江河湖海水脉交汇，奠定了今日上海城市空间的基础格局。"城市肌理"展项重点展示城市自然肌理（海岸线演变、水系演变）和城市人文肌理（行政区划演变、道路发展演变、建筑肌理形态）。

城市自然肌理呈现了上海成陆的复杂自然演变和逐步奠定的海岸线轮廓演进，以及分片治理、人工调控的河网水系演进过程。

城市人文肌理梳理出上海历次行政区划调整，清晰展示上海历次重大变革对经济社会发展的推动与促进，呈现上海道路发展演变对于城市形态发展的重要意义，显示出上海日渐完善的城市道路在方便居民出行、助力城市交通的同时，也塑造着城市的特色风貌、见证着城市的发展历程。上海的建筑肌理形态则在各个时期不同空间形态的圈层拓展，彰显出属于各自时代的价值取向和审美风格。

Content

Over the past millennium, the evolution of rivers, lakes and seas has laid the foundation of Shanghai's urban contour today. This area focuses on the city's natural fabric—the evolution of its coastline and water system--and human-made fabric--the evolution of administrative districts, road network and architectural profile.

The natural fabric part deals with the complex process of land and coastline formation Shanghai has undergone, and the evolution of its river network and water system through governance division and artificial control.

The part about the city's human-made fabric chronicles the successive changes to Shanghai's administrative districts. It shows how these changes fueled the economic and social development and why the evolution of Shanghai's road network is critical to the development of urban scape. It tells visitors how the ever-improving urban road network facilitates residents' travel and traffic, and how it shapes the city's profile and witnesses its growth. Shanghai's architectural profile has expanded in different spatial forms in different periods, revealing the values and aesthetics pertaining to each particular era.

展项解读

"人文之城"的视觉设计基础元素为圆形,"城市肌理"亦选用圆形为母题设计展台、地屏、肌理墙等展陈家具。展项主体包括一面弧形展墙、两组弧形展台,以及由三个不同尺寸的圆形有机拼接形成的花瓣状发光二极管(LED)地屏。展墙、展台向内环绕包裹着地屏,通过地屏相互衔接,三者形成"高、中、低"三个错落有致的观赏层级。

"锦绣上海"顾绣肌理墙、花瓣形的多媒体地屏演绎,结合弧形查询台,以艺术化的方式勾勒出上海在时间、空间和人文活动共同作用下的城市演进过程。为拓展观众认知,展项基于专业的规划视角引入上海的海岸线演变、水系演变、行政区划演变、道路发展演变、建筑肌理形态等历史地图、文档史料,观众通过点击、滑动操作与五块显示屏实时互动,深入认知上海城市空间结构的多次演变过程,沉浸式体验上海城市发展的脉络。

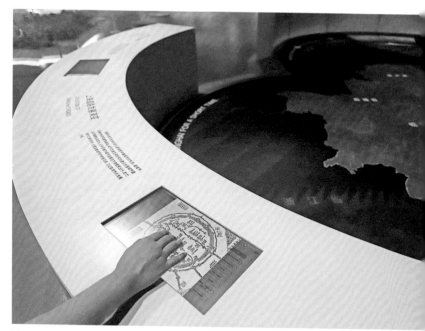

弧形展台内嵌互动触摸屏,观众可以通过点击、滑动操作与屏幕实时互动
Visitors can click and move their fingers on the five display screens to learn more

Interpretation

The design of the "Humanistic city" gallery pivots in full circle as the visual element, while the "Urban Fabric" area also chooses circle as the motif, which can be found in the stands, floor screens and the texture wall. The main structure of the exhibition consists of a curving wall, two circular-shaped display stands, and a petal-shaped LED floor screen formed by three different circles in different sizes. The exhibition wall and stands encompass the floor screen while the three are connected to each other through the latter, forming a high, medium and low level system.

The embroidery-textured wall, the petal-shaped multimedia floor screen, and the curving information desk artistically outline Shanghai's urban evolution with the joint involvement of time, space and human activities. To create the awareness of the evolution of Shanghai's coastline, water system, administrative districts, road network, and architecture, the exhibition area introduces historical maps and documents needed for professional planning. Visitors can click and move their fingers on the five display screens to trace the evolution of Shanghai's urban spatial contour and feel the pulse of Shanghai's urban development.

技术特色

材质与工艺的创新 ——《锦绣上海》顾绣肌理墙

在"城市肌理"展项设计中，城市肌理展墙以《锦绣上海》为设计主题，以顾绣作为表现手法。顾绣作为上海的国家非物质文化遗产，自明嘉靖年间发展至今已有四百余年，其工艺将宋绣中传统的针法与国画笔法相结合，针法复杂多变，艺术性极高。在展项中，顾绣展墙以上海城市肌理为蓝本，经由顾绣传人绣制转译，以针代笔、以线代墨，以精湛技法，历经半年时间绣制出六片绣样，缝合上海不同区域的肌理，展现出一幅气韵生动的城市空间景象。

绣品以传统国画拓印装裱工艺——布面拓印宣纸的双层装裱方式，将绣品叠合拓印于底部钢化玻璃表面，形成长 7 米、高 2.5 米的巨幅城市肌理画卷。顾绣肌理墙上叠加数字投影，依次展示了中共一大会址及周边区域、外滩历史文化风貌区、衡山路 - 复兴路历史文化风貌区、朱家角历史文化风貌区、陆家嘴金融贸易区、临港新片区等区域肌理。传统文化与现代科技的美丽"邂逅"，使上海丰富的空间格局发展、富有差异性的肌理演变"活"了起来，使观众通过沉浸式的体验与城市故事产生共情。

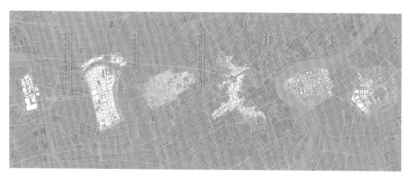

顾绣肌理墙结合数字投影，勾勒出一幅城市肌理长卷
Gu embroidered wall and digital projection showing Shanghai's changes

Technical Features

Technical Innovation: *Gu* embroidery titled "Magnificent Shanghai"

In the "Urban Fabric" area, the exhibition wall is striking for its tapestry of *Gu* embroidery titled "Magnificent Shanghai." As an item of national intangible cultural heritage of Shanghai, *Gu* embroidery has a history of four centuries. The well-trained artisans tapped into the traditional stitches of Song embroidery and the brushwork of Chinese painting. The city's urban fabric is finely crafted with stitches and threads – rather than brush strokes and lines--on six pieces, each representing a part of Shanghai. They spent six months on this remarkable artifact.

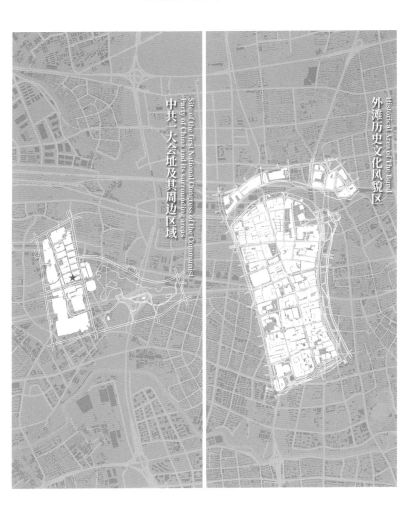

中共一大会址及其周边区域
Site of the first National Congress of the Communist Party of China and its surrounding areas

外滩历史文化风貌区
Historical Area of The Bund

The embroidery works, 7-meter wide by 2.5-meter high, is superimposed and rubbed on tempered glass surface, via the cloth-rubbed-rice paper framing way frequently used in the field of traditional Chinese painting. The digital projection superimposing on the *Gu* embroidered wall, successively shows the site of the First CPC National Congress, its surrounding landscape, the Bund Historical Cultural Areas, Hengshan Road-Fuxing Road Historical Cultural Areas, Zhujiajiao Historical Cultural Areas, Lujiazui, Lingang and others in detail. When traditional culture meets modern technologies, Shanghai's flexible spatial development and differentiated texture evolution becomes alive, allowing the visitors strong empathy with Shanghai's changes.

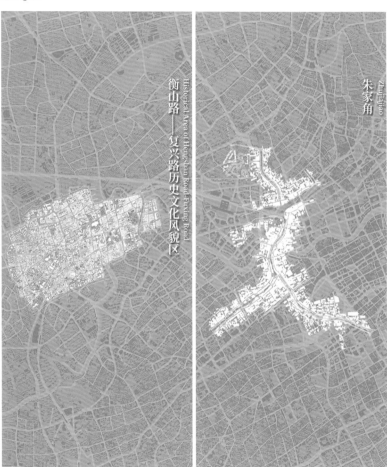

衡山路——复兴路历史文化风貌区
Historical Area of Hengshan Road-Fuxing Road

朱家角
Zhujiajiao

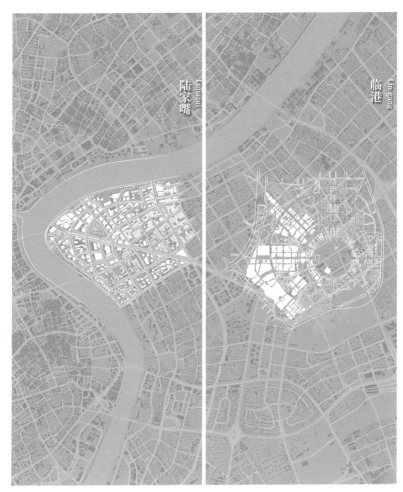

陆家嘴 Lujiazui

临港 Lin-gang

花瓣形 LED 地屏极具巧思，所展示的《江海之韵》《都市脉络》《大地年轮》三部影片基于地屏形态作出适应性设计，将展墙、展台与地屏连接为视觉统一体，极具艺术性。《江海之韵——用力量冲刷城市轮廓》以流动的线条勾勒上海的海岸线变迁，海岸分层肌理与蔚蓝色的海水相交叠，幻化出美丽的视觉景象。《都市脉络——用线条勾勒城市肌理》采用叶脉隐喻上海城市道路的发展和演变。影片开篇以叶脉肌理，由屏幕中央缓缓蔓延至四周，用叶脉生长喻意城市道路的生长。《大地年轮——用时间标记城市印记》图形灵感来自树木的年轮，上海城市的演变、城市区划的演进如同年轮一般缓缓蔓延与聚合。

地屏选用像素点间距仅有 1.5 毫米的 LED 显示屏，分辨率高，同时采用优化图像处理算法，展示出高清画质和细腻色彩，为观众带来精细的视觉体验。其异形花瓣形态也为 LED 模块的设计带来挑战，设计一改 LED 屏常用的正方形灯珠模组，遵循矩阵灯点的布置切割 LED 模组，保留灯珠矩阵完整度及识别位置。

Floor screen: An innovative display device (that showcases technology)

The petal-shaped LED floor screen is a must-stop: Here you can see a three-part video, "Sea Moods," "City Fabric" and "The Wheel of the Earth," that is adapted to the shape of the screen. The footage is cascaded on the floor screen, display wall and stands but the view appears uninterrupted. "Sea Moods" uses flowing lines to depict Shanghai's changing coastline; the mix of geological layers of the coastal area with azure seawater delivers a stunning vision. "City Fabric – Sketch the City in Lines," starts with a leaf vein that fans out from the center to the limits and takes the growth of leaf veins as a visual metaphor for the sprawl of urban traffic network. The inspiration for "The Wheels of the Earth" comes from a tree's annual wheels, which symbolizes the expanding city and its zoning changes.

The ground screen uses high-resolution LED displays with a pixel pitch of only 1.5 mm; with the benefit of optimized image processing algorithms, it delivers high-definition pictures in delicate color shades. Its "petals" in diverse sizes also poses a challenge for the design of the LED module. The design has given up the square light-emitting diode module commonly used in LED screens in favor of the modules following the arrangement of matrix light dots cutting LED modules, retaining the integrity of the LED matrix and their identification positions.

《大地年轮》在花瓣形地砖屏内不断变化，展现出上海城市区划的演变
"The Wheels of the Earth": The kaleidoscopic images in the petal-shaped floor screen trace the evolution of Shanghai's urban development and zoning.

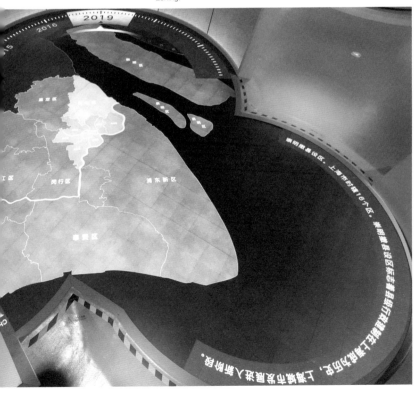

风貌格局 Feature Patterns

"风貌格局"展项轴测图
Axonometric drawing of "Feature Patterns" exhibit area

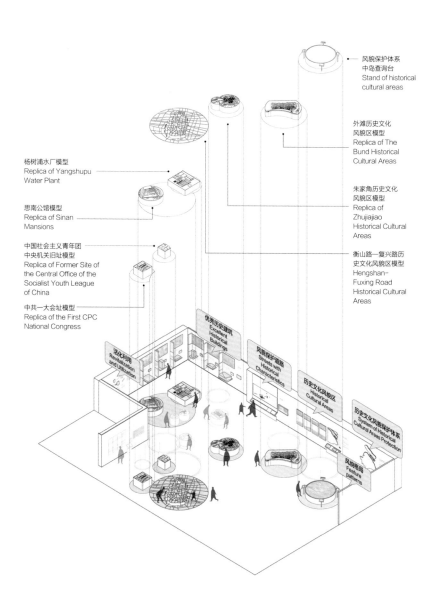

风貌保护体系
中岛查询台
Stand of historical
cultural areas

外滩历史文化
风貌区模型
Replica of The
Bund Historical
Cultural Areas

杨树浦水厂模型
Replica of Yangshupu
Water Plant

朱家角历史文化
风貌区模型
Replica of
Zhujiajiao
Historical Cultural
Areas

思南公馆模型
Replica of Sinan
Mansions

中国社会主义青年团
中央机关旧址模型
Replica of Former Site of
the Central Office of the
Socialist Youth League
of China

衡山路—复兴路历
史文化风貌区模型
Hengshan–
Fuxing Road
Historical Cultural
Areas

中共一大会址模型
Replica of the First CPC
National Congress

优秀历史建筑
Excellent
Historical
Buildings

活化利用
Revitalization
and Utilization

风貌保护道路
Streets with
Historical
Characteristics

历史文化风貌区
Historical
Cultural Areas

历史文化风貌保护体系
System of Historical
Cultural Areas Protection

风貌格局
Feature
patterns

展示内容

自 2003 年 1 月《上海市历史文化风貌区和优秀历史建筑保护条例》实施以来,上海积极推进历史文化风貌区和优秀历史建筑的保护工作。2019 年,上海通过对该条例的修订,进一步贯彻落实中央和市委关于历史文化遗产保护的最新要求和部署,践行"整体保护、以用促保"的理念,聚焦"扩大保护范围、强化政府责任、完善保护措施、促进活化利用"等方面工作。

"风貌格局"立足建成既有深厚历史文化底蕴、又有鲜明时代特征的国家历史文化名城的目标,讲述上海在历史文化风貌区、风貌保护道路、优秀历史建筑方面的保护举措与实践,以及对其进行活化利用的探索。

"历史文化风貌区"展示了人民广场、老城厢、衡山路 – 复兴路历史文化风貌区和金泽、新场、朱家角历史文化风貌区等典型案例。"风貌保护道路"展示了上海 64 条永不拓宽的风貌保护道路格局。"优秀历史建筑"展示了上海具有代表性的优秀历史建筑的基本概况、建筑图纸及相关影像。"活化利用"展示了上生新所、思南公馆、外滩源等城市更新重点项目。

"风貌格局"展项实景
"Feature Patterns" exhibit area

Content

Since the "Regulations of Shanghai Municipality on the Protection of The Areas with Historical Cultural Features and the Excellent Historical Buildings" were announced in January 2003, Shanghai has been pressing on the job seriously. The 2019 revised version confirms the latest requirements and deployments of the central and municipal governments on the protection of historical and cultural heritage. Under its guidance, Shanghai will further implement the policy of "overall protection and preservation through utilization." The holdings show what make the city thick: expanding the scope of preservation, strengthening government responsibility, improving preservation measures and promoting utilization.

The "Feature Patterns" exhibit area illustrates Shanghai's crave for a culture-rich city with multiple heritage resources and contemporary features. It surveys Shanghai's efforts in protecting its historical cultural areas, roads and heritage buildings, and the revitalization and utilization of the latter.

The "Historical Cultural Areas" part gives an in-depth look at the cases of People's Square, the old Shanghai Town, the Hengshan-Fuxing Road neighborhood in the city center and the tradition-rich Jinze, Xinchang and Zhujiajiao in the suburbs. The "Roads in Historical Cultural Areas" section tells the information about the 64 roads under municipal protection. The "Excellent Historical Buildings" section delivers an overview along with architectural drawings and related images. The "Revitalization and Utilization" area allows the visitors to step into such landmarks as the Columbia Circle, Sinan Mansions and Waitanyuan.

上海市域历史文化保护规划示意图
Diagram of Shanghai Historical and Cultural Heritage Preservation Plan

上海市域历史文化保护规划示意图
Diagram of Shanghai Historical and Cultural Heritage Preservation Planning

N

图例
历史城区   历史文化名镇   省名界
中国历史文化名镇   全域
中国历史文化名村   市干线网

朱家角历史文化风貌区
Zhujiajiao Historical Cultural Area

建国西路
West Jianguo Road

武康大楼
Wukang Building (I.S.S Normandie Apartment)

展项解读

　　"风貌格局"展项延续"城市肌理"的时空叙事，
以层叠的历史厚度展现上海的发展脉络。为提供可供
观众零距离观展的空间，展项采用大幅面展墙结合环
绕式展台的布局方式，营造开放式的观展空间。灵活
布局且形态各异的模型展台是展项视觉设计的重点。
展项设计突出圆形的主题元素，其中氛围灯、地贴及
展陈家具均采用弧形、环形等圆形为母题的形态要素。
重点案例利用大、中、小不同体量、不同层级的模型
进行展示，并采用圆形、方形、不规则形等多种形态，
提升观展的趣味性。

展区模型选用原木材质，具有较强可塑性的原木经过切割、雕刻、车削、热弯等加工方式呈现出各式造型。自然原木肌理所蕴含的生命质感与清新的纹理产生独特的质感，增强了空间的亲和力，让严谨、系统的建筑模型有了温度。

Interpretation

Picking up the spatial narrative of "Urban Fabric," the "Feature Patterns" area traces the evolution of Shanghai. The set-up, featuring a large exhibition wall combined with a wrap-around stand, makes for an open space. The display stands--wrought in different shapes and placed in an unusual layout--highlight the effort of designers, who introduces plenty of circles to the space. Ambience lights, floor stickers and furniture pieces are all shaped in arcs, rings and other variations of circles. Major cases of heritage preservation are presented at stands in diverse sizes and at staggered levels. The juxtaposition of various geometrical shapes, i.e., circles, squares and irregular shapes, seeks to catch your fancy.

Pieces of wood, the chosen material for the replicas of buildings, assume various shapes after they undergo a series of processing procedures, cutting, carving, turning and heat-bending. The lovely natural texture of wood echoes with the vitality of life and produces a unique feel. The material of choice enhances the vibrancy of space and lends a human touch to the otherwise lifeless replicas.

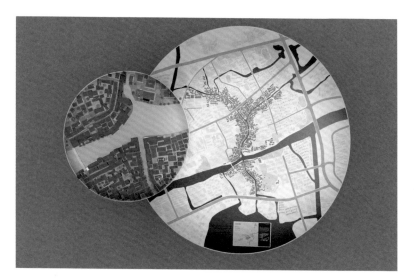

朱家角历史文化风貌区模型设计图
Tableau of Zhujiajiao Historical Cultural Areas

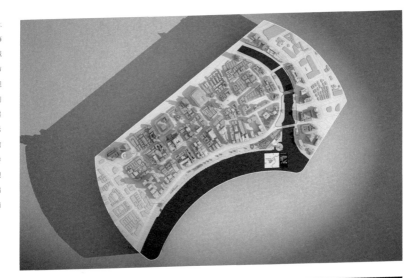

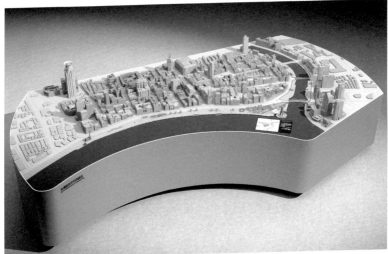

外滩历史文化风貌区模型设计图
Tableau of Bund Historical Cultural Areas

为进一步打造沉浸式观展空间，展项采用不同展陈媒介，将数字影像与逐层投影融入展品中。《梧桐树下的故事》氛围投影和衡山路—复兴路历史文化风貌区模型的结合，使观众感受到整个片区的风貌与格调。影片采用"梧桐深处""信步街巷""衡复焕新"三个核心篇章和梧桐叶这个经典元素，顺延本层展陈"面—线—点"的叙述逻辑，逐层投影结合大区域俯瞰视角，勾连起观众对于上海的城市记忆，隐喻衡复的故事永不停歇。

"风貌保护道路"影像由中国摄影家协会会员、上海摄影家协会理事郑宪章先生组织拍摄完成。从传统的道路中央切片式拍摄方式转向反映风貌道路两侧第一界面建筑、景观风貌与日常生活的视角，画面如同城市山水长卷缓缓展开，让观众近距离感受城市的温度。影像采用定制长条屏的呈现方式。

衡复历史文化风貌区
Hengshan Road-Fuxing Road
Historical Cultural Areas

"梧桐树叶"在衡复风貌区模型上缓缓飘落
Plane tree leaves falling over the tableau of the Hengshan-Fuxing Historical Cultural Areas

To enhance an immersive viewing experience, the exhibit area uses different display media to incorporate digital images and layer-by-layer projections into the display. The video "The Story under the Plain Trees," and the tableau of the Hengshan-Fuxing Road Historical Cultural Areas combine to allow visitors the look and feel of the neighborhood. The three parts of the video, "Tucked Deep in the Foliage," "Strolling the Lane" and "Renaissance of the Neighborhood," and the classic element of plane tree leaves follow the narrative logic of "surface-line-point." They are compounded by the layer-by-layer projection of aerial views, reviving the memories of Shanghai and perpetuating the legend of the leafy neighborhood.

The photos of "Streets with Historical Characteristics" are works by a team led by Mr. Zheng Xianzhang, a member of the China Photographers' Association and a director of the Shanghai Photographers' Association. Rather than symmetrical shots from the center of the roads, these photos capture the architecture, road capes and plenty of life on both sides. Like a long scroll of urban landscape unfolded, they give you an immediate impression of the cityscape. The images are presented on custom-made oblong screens.

64 条永不拓宽的风貌保护道路分布图
Distribution map of the 64 streets with historical characteristics that will never be expanded

桃江路沿街面
Front view of the Taojiang Road

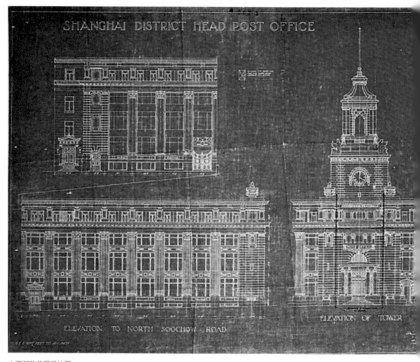

上海邮政总局设计图
A drawing of Shanghai General Post Office Building

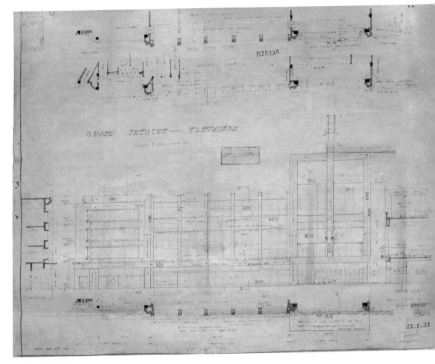

大光明电影院设计图
A drawing of the Grand Theatre

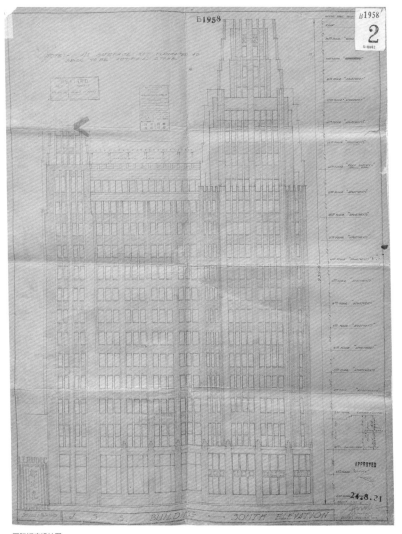

国际饭店设计图
A drawing of the Park Hotel

杨浦图书馆旧影
An old photo of the Yangpu Library

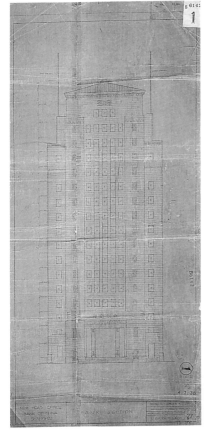

中国银行大楼设计图
A drawing of the Bank of China Building

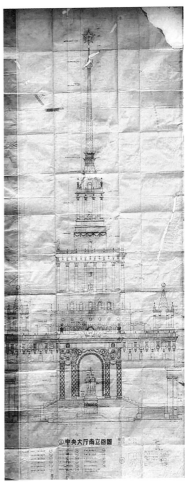

上海展览中心设计图
A drawing of the Shanghai Exhibition Centre

"活化利用"设有一块氛围投影区，结合一幕幕场景切换，勾勒出老建筑"活化"焕发新活力的景象。

An exciting assemblage of images shown in the projection area illustrates the renaissance of old buildings through revitalization and utilization.

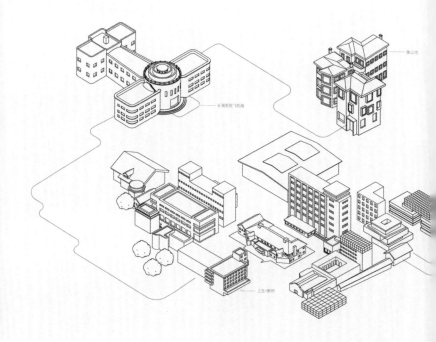

"活化利用"墙面投影
Projection on the wall

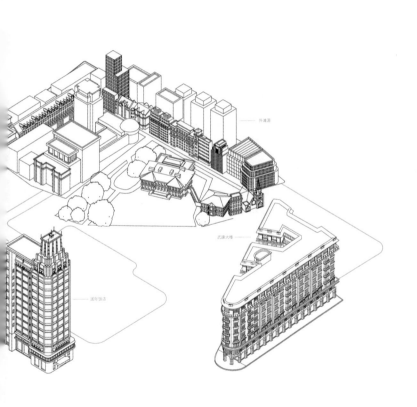

海军俱乐部外立面保护更新前与保护更新后
Past and Now: revitalization and utilization of the Navy Club

19 世纪末期及 1950 年代的外滩源
Waitanyuan at the end of 1800s and 1950s

技术特色

　　外滩历史文化风貌区实体模型运用 AR 增强现实技术，将数字内容增强到真实观展世界，实现信息无缝集成。虚拟影像中的光影景致和动态车流增加了数字场景的真实感，观众可通过展厅的平板电脑（iPad）切换日夜场景，与历史风貌区实体模型进行互动体验。

Technical Features

The replica of the Bund Historical Cultural Areas taps into AR technology to achieve seamless integration of digital and physical replicas. The scenes and dynamic traffic flow in the virtual image make the digital scene more lifelike, and visitors can choose the day and night scenes through the iPads, come into dialogue with the tableau of Historical cultural areas in the exhibition hall.

展项运用 AR 技术，实现数字模型与实体模型的无缝集成
The exhibit area uses AR technology to propel seamless integration of digital and physical replicas

城市更新　Urban Regeneration

"城市更新"展项轴测图
Axonometric drawing of "Urban Regeneration"

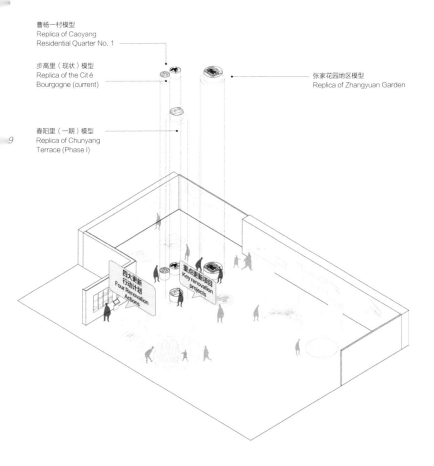

曹杨一村模型
Replica of Caoyang
Residential Quarter No. 1

步高里（现状）模型
Replica of the Cité
Bourgogne (current)

张家花园地区模型
Replica of Zhangyuan Garden

春阳里（一期）模型
Replica of Chunyang
Terrace (Phase I)

四大更新
行动计划
Four Renovation
Actions

重点更新项目
Key renovation
projects

展示内容

　　上海积极推进城市更新工作。"城市更新"展项重点展示上海在政策法规制定、更新实践推进、实施机制创新等方面的积极探索，聚焦上海 2016 年以来开展的"共享社区计划""创新园区计划""魅力风貌计划""休闲网络计划"四大更新行动计划的具体实践，通过展示张园、曹杨一村、春阳里、步高里等重点更新项目，体现历史建筑的保护和居民生活品质的提升。

Content

Shanghai keeps pressing forward its campaign of urban regeneration. The eponymous exhibition area draws attention to the city's efforts: the formulation of policies and regulations, practices of urban regeneration, and innovation of implementation mechanisms. It shows vividly the Shared Community Plan, Creative Park Plan, Charming Landscape Plan, and the Leisure Network Plan that Shanghai has launched since 2016. You will be impressed by the spruce-up of the Zhangyuan Garden, Caoyang Residential Quarter No. 1, Chunyang Terrace and the Cité Bourgogne, how heritage architecture is preserved and residents' quality of life has improved.

杨浦滨江
Yangpu section of Huangpu waterfront

展项解读

　　"城市更新"展项通过图板、触摸屏、图书三者的有机结合，让观众对城市更新有全方位的感知和多角度的知识拓展。案例模型为展项内容作出更具深度的实体延展，如张园模型以 1:200 的比例细致呈现出内部功能活化后完整保留的石库门原型。步高里模型采用 1:150 比例更好地展示了里弄建筑更新后重现的历史风貌，反映居民生活质量的显著提升。

《有机更新四大计划》封面
Cover of *Four Plans of Shanghai Organic Urban Renewal*

Interpretation

The exhibit area of "Urban Regeneration" in combination of boards, touch screens and books allows the visitors more knowledge of urban renewal from multi perspectives. The replicas make more stereoscopic images. For example, the replica of Zhangyuan Garden at a scale of 1:200 meticulously presents the prototype of Shikumen well preserved with flexible and convenient functions; The Lilong Bourgogne replica at a scale of 1:150 shows its original style after preservation-oriented renovation, and reflects the significant improvement of the residents' life.

张园模型局部
Replica of Zhangyuan Garden

步高里模型展示里弄建筑风貌
Replica of Lilong Bourgogne

黄浦江、苏州河：
迈向具有全球影响力的世界级滨水区
The River and the Creek

"一江一河"展项轴测图
Axonometric drawing of "The River and the Creek"

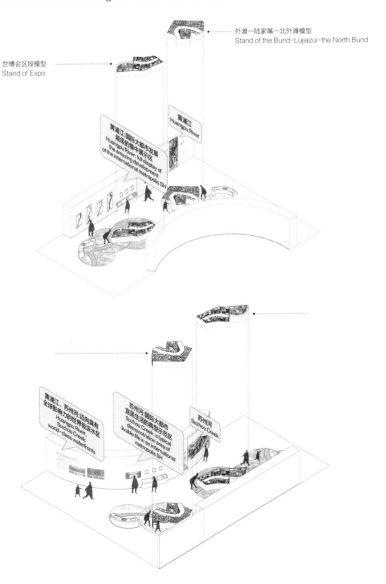

外滩—陆家嘴—北外滩模型
Stand of the Bund–Lujiazui–the North Bund

世博会区段模型
Stand of Expo

黄浦江
Huangpu River

黄浦江·国际大都市发展
能级的集中展示区
Huangpu River: full display of
the amazing development
of the international metropolis SH

黄浦江、苏州河:迈向具有
全球影响力的世界级滨水区
Huangpu River,
Suzhou Creek:
world-class waterfronts

苏州河·国际大都市
宜居生活的典型示范区
Suzhou Creek: a typical
demonstration area of
livable life in an international
metropolis SH

苏州河
Suzhou Creek

展示内容

　　黄浦江与苏州河（以下简称"一江一河"）是上海建设"国际大都市"的代表性空间和标志性载体。展项围绕建设"具有全球影响力的世界级滨水区"的规划目标，布局"黄浦江：国际大都市发展能级的集中展示区"和"苏州河：国际大都市宜居生活的典型示范区"两大板块。展项从"回望""开发历程""规划概况""规划目标""规划策略""功能分区""地区案例""公共空间贯通"等专题对黄浦江、苏州河展开阐述。

　　展区设有两大模型展台，由外滩、虹口北外滩及浦东陆家嘴组合的模型，呈现黄浦江沿岸三个核心区块——陆家嘴金融城、外滩金融集聚带和北外滩所构成的上海发展的"黄金三角"整体天际轮廓；浦东世博会区段模型，呈现了在历史、现在和未来的交汇下，两岸联动、交相辉映的美好景象。

Content

10

The waterfronts of the Huangpu River and the Suzhou Creek are the city's iconic spaces. The reinvigoration campaign of the River and the Creek reflects Shanghai's ambition to gain a global distinction as a city with superb waterfront. The exhibition area covers the planning goal of building world-class waterfront. It comprises of two sections: "Huangpu River – A centralized display of development capabilities in an international metropolis" and "Suzhou Creek – A typical demonstration zone of livability in an international metropolis." Each section gives an extensive survey on "Retrospect," "Development History," "Planning Overview," "Planning Goal," "Planning Strategy," "Functional Zoning," "Case Study," and "Reinvigoration Campaign."

In this area there are two stands. One features the tableau of the Bund – the North Bund in Hongkou – Lujiazui in Pudong. Here the three landmarks along the Huangpu River – Lujiazui Financial City, the Bund Financial Cluster and the North Bund – constitute a dynamic skyline. The other, the tableau of the Pudong section of the Expo site affords a wonderful view of both sides of the river; their past, present and future are equally glorious.

05

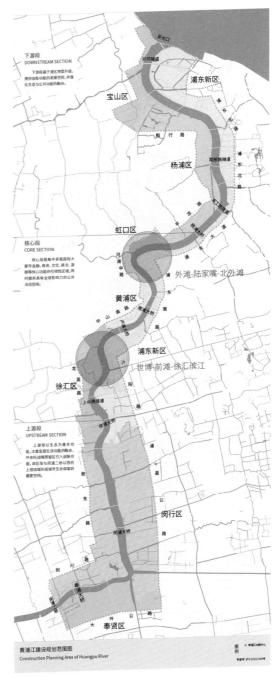

宝山区

浦东新区

吴淞口

外环隧道

殷行路

翔殷路隧道

杨浦区

虹口区

河南中路

外滩-陆家嘴-北外滩

黄浦区

中山南路

南浦大桥

浦东新区

龙吴路

世博-前滩-徐汇滨江

徐汇区

上中路隧道

徐浦大桥

闵浦大桥

闵行区

剑川路

奉贤区

闵浦二桥

叶榭公路

黄浦江建设规划范围图
Construction plan of Huangpu
River

黄浦江建设规划范围图
Construction Planning Area of Huangpu River

图例

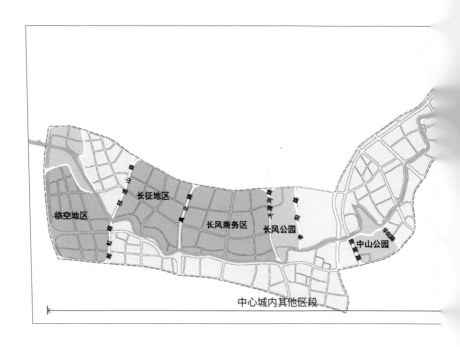

临空地区　长征地区　长风商务区　长风公园　中山公园

中心城内其他区段

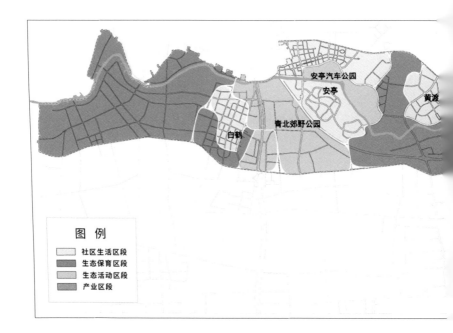

安亭汽车公园　安亭　黄渡　青北郊野公园　白鹤

图　例

社区生活区段
生态保育区段
生态活动区段
产业区段

苏州河沿岸外环内（内环内东段和中心城内其他区段）功能布局图
Function layout map of the Suzhou Creek within the outer ring (the eastern section within the inner ring and other sections in the central urban area)

不夜城

苏河湾

外滩源

东斯文里

图　例

公共活动区段
商务区段
社区生活区段
生活活动区段

内环内东段

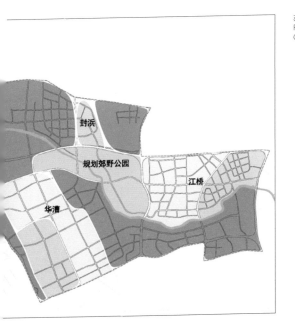

苏州河外环外区段功能布局图
Function layout map of the Suzhou Creek outside the outer ring

封浜

规划郊野公园

江桥

华漕

展项解读

　　展项应用模型、多媒体、抽拉阅读等多种展示形式，营造出沉浸式动态观展空间，渲染出黄浦江、苏州河与城市对话的亲切氛围。

　　黄浦江区段模型采用1:1000比例，结合原木材质，经匠人手工制造，展现出机械雕刻无法企及的细腻质感，凸显展项的人文温度。模型中引入的动态抽拉板组件，有效延展了展台观赏的丰富性与场景感。展台底部铺设了最新地理影像圆形的地贴，地贴影像与实体模型融为一体，将黄浦江的空间更加完整细致地展现在观众面前，为观众认知黄浦江带来更加直观的三维立体感受。

Interpretation

The exhibit area creates an immersive, dynamic viewing experience with replicas, multimedia and pull-out reading devices, presenting a compelling look at the Huangpu River, the Suzhou Creek and beyond.

The 1:1000 scale replica of the Huangpu River waterfront takes wood as the material. The hand-made craftsmanship shows a delicate texture that no mechanical carving can match, suggesting a human touch of the exhibition. The introduction of dynamic pull-out panels to the replica lends a poignant, immersive feel to the visit. Around the bottom of the display stands are circular floor stickers that sport the latest map of the waterfront. The stickers and the replicas combined are designed to be a fascinating showcase of the waterfront. The 3D rendering presents an immediate image of the mighty river.

外滩—陆家嘴—北外滩模型
Replica of the Bund, Lujiazui, and the North Bund

浦东世博会区段模型
Replica of the Pudong section of Expo site

　　苏州河版块应用壁挂多媒体装置——三维（3D）浮雕模型叠加投影，从空间层面丰富了展陈形式，带来更具活力的观展体验。

　　为丰富观展层次，除《黄浦江》《苏州河》主题影片展示外，展区内的触摸屏多维立体呈现"一江一河"的规划策略、地区案例、历史影像、文献史料等，为观众建构出一幅关于"一江一河"的记忆叠层。

The Suzhou Creek is wall-mounted multimedia installations - three-dimensional (3D) relief replicas overlaid with projections - to enrich the exhibition form from the spatial level and bring a more dynamic display experience.

In addition to the documentaries of Huangpu River and Suzhou Creek, a touch screen delivers a sea of information on the two waterways in a multi-dimensional manner: the planning strategy, regional case studies, historical images, and archived documents and materials. It truly has something special for everyone.

苏州河今昔对比
Suzhou Creek now and then

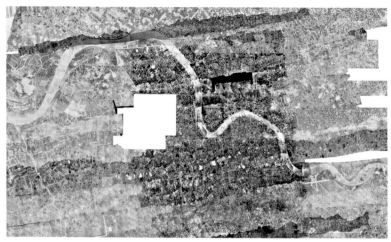

1948

1979

2010

11

2017

2020

城市设计标准　**Urban Design Standards**

"城市设计标准"展项轴测图
Axonometric drawing of "Urban Design Standards"

1 《城市设计的管控方法——上海市控制性详细规划附加图则的实践》
The Control Methods - Practice of Detailed Shanghai Control Planning with Pictures

2 《上海市街道设计导则》
Shanghai Street Design Guidelines

3 《上海市河道规划设计导则》
Shanghai River Planning and Design Guidelines.

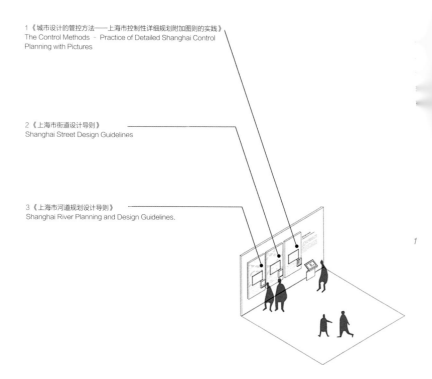

1

展示内容

　　上海坚持结合实际，对标国际标准，在以人为本的城市设计理念指导下，以管控体系贯穿规划建设全流程，逐渐形成一套具有上海特色的城市设计管理机制和方法。"城市设计标准"展项重点展示《城市设计的管控方法——上海市控制性详细规划附加图则的实践》《上海市街道设计导则》《上海市河道规划设计导则》等实践成果，体现城市在精细化管理方面的创新举措。

Content

Shanghai has developed its urban design guidelines and methods while referring to its reality and international standards. Observing the principle of "human-centric urban design," the city uses a control system throughout the whole process of planning and construction. The "Urban Design Standards" exhibit area highlights its achievements, such as the *Control Methods – Practice of Detailed Shanghai Control Planning with Pictures*, the

《城市设计的管控方法——
上海市控制性详细规划附加图则的实践》
SHANGHAI URBAN DESIGN MANAGEMENT –
REGULATORY ADDITIONAL PLAN PRACTICE

　　自2011年开始，上海推行城市设计管控制度创新方法，即城市设计法定化成果——附加图则。附加图则明确以城市公共空间为核心管控对象，归纳出建筑形态、开放空间、交通空间、功能空间和历史风貌五种空间类型，以"工具箱"的形式统一设计管控的表达。通过虹桥商务区、世博、前滩、徐汇滨江、临港、桃浦等地区的具体实践，不断提升城市空间形象和品质。

城市设计标准展项
"The Urban Design Standards"
exhibit area

Shanghai Street Design Guidelines and the *Shanghai River Planning and Design Guidelines*. All these codes reflect the city's innovations in refinement management.

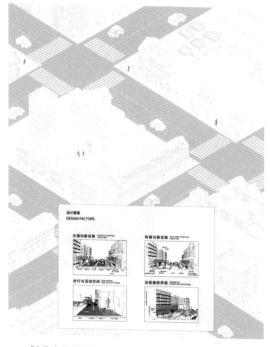

城市设计标准展项
The Urban Design
Standards exhibit area

《上海市街道设计导则》
SHANGHAI STREET
DESIGN GUIDELINES

　　设计导则围绕安全、绿色、活力、智慧的目标,对街道空间内的交通功能设施、步行与活动空间、附属功能设施和沿街建筑界面四大要素进行设计引导,帮助街道的设计者、建设者、管理者和使用者从更广阔视角认识街道,用多元手段塑造承载市民生活、沉淀市民情感的魅力空间。

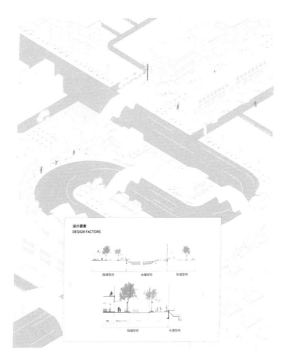

《上海市河道规划设计导则》
SHANGHAI RIVER PLANNING
AND DESIGN GUIDELINES

《上海市河道规划设计导则》作为全国同类导则先河,立足"生态为先、安全为重、人民为本、文化为魂"的规划思路,强调"水陆统筹、水岸联动、水绿交融、水田交错",全过程指导市域范围内的河道及滨河空间的规划、设计、建设、管理和运维,为各级管理者提供技术和管理支撑,为设计师提供设计指引,为市民提供更多水清岸绿、丰富多样、活力四射的滨水空间。

展项解读

　　"城市设计标准"展项采用二维数字手绘方式,绘画内容选自三本图书中的核心要素,将城市内部道路、水系及绿化等元素置于数字网格中,形成生动细腻的轴测图,并配合实体图书的展示,与观众亲切互动,观众可直观地了解上海城市精细化管理的特点以及贯穿全流程的管控体系。

Interpretation

This area features digital hand-crafted drawings selected from the three aforementioned codes. Placing the city's internal roads, drainage systems and green spaces in a digital grid, it presents them in a finely crafted axonometric drawing; along with the display of books, the drawing illustrates the characteristics of Shanghai's urban refinement management and the all-inclusive control system.

观众点击触摸屏，可以观看三本导则的详细内容
Visitors can read through the three codes on the touch screen

15 分钟社区生活圈
15-minute Community Life Circle

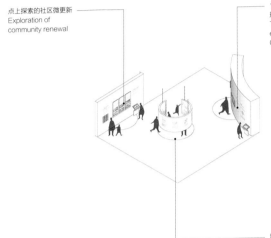

点上探索的社区微更新
Exploration of
community renewal

15 分钟社区生活圈
规划理念与实践探索
Theoretical and practical
exploration of 15-minute
Community Life Circle

新华路街道
"15 分钟社区生活圈"行动规划
Action Plan of Xinhua Road Sub-
district 15-minute Community Life
Circle

展示内容

　　社区是城市生产、生活和治理的基本单元，是服务市民和基层治理的"最后一公里"。"15 分钟社区生活圈"展示上海积极开展社区更新探索，重点聚焦社区微更新、"15 分钟社区生活圈"社区更新行动等案例，反映上海基于城市最小单元的创新实践，全面提升城市品质的共建共治共享行动，为推动实现"人民城市"的持续建设提供示范性的城市样板，贡献上海智慧。

　　展项将社区更新行动首批试点区域之一——新华路街道作为展示案例，展现新华社区坚持以人民为中心，强化顶层设计，鼓励多元参与，用"社区生活圈"绘制出一个愿景（花园社区、人文新华）、五大目标（宜居、宜业、宜游、宜学、宜养）及五类行动（安居行动、乐业行动、漫游行动、趣学行动、怡养行动）的"新华蓝图"，打造人人参与构建社区的"新华样本"。

Content

Communities are the basic units of urban production, life and governance, and they are the barometers of the "last mile" citizen service and grassroots governance. The "15-minute Community Life Circle" area is devoted to the cases Shanghai has done in community renewal, reflecting the city's innovative practices based on the smallest urban units and its comprehensive efforts in urban quality improvement. They provide exemplary replicas and highlight Shanghai's commitment to being a "People's City."

The exhibition cites Xinhua Road Sub-district, one of the first pilot areas of the Community Renewal Initiative, as a case to show what the community has done: It sees people's needs as its top concern, it rolls out top-level design, encourages the engagement of multiple parties, and propose its vision—a leafy, culture-rich community—five goals, e.g., livable, workable, playable, learnable, and age-friendly, and initiated five campaigns. They create a laudable Xinhua Blueprint: Everyone can participate in the community-building.

15 分钟生活圈社区设施圈层布局示意图
Diagram of facilities layout in a 15-minute Community Life Circle

展项解读

　　"15分钟社区生活圈"展项以中央环形投影区呼应"社区生活圈"概念,自然连接左右两侧展墙,构成开敞灵动的展示空间。影片以数字手绘长卷为底,记录新华路街道的日常生活。长卷内嵌入五块分镜头,画面分别展示了新华路街道的五大行动,构成一幅生动的生活圈图景。

Interpretation

This area echoes the concept of "community life circle" with a central circular projection space, which naturally links the left and right walls to form an open and dynamic display space. The film uses a digital hand-drawn scroll as a background to reflect the daily life of Xinhua Road. Five sub-cameras are embedded in the scroll, displaying the five campaigns of the Xinhua Road Sub-district - truly a vivid picture of the life circle.

技术特色

　　投影区采用影像融合技术,三台超短焦投影设备在小型环状空间内巧妙布局,实现在小直径圆形空间内最佳的投射效果,使影像得以完整清晰呈现。

Technical Features

The projection area uses image fusion technology, with three ultra-short-throw projection devices cleverly laid out in a small circular space to achieve the best projection effect in a small-diameter circular space, allowing images to be presented completely and clearly.

中央环形投影区的创意性设计
The ring projection area

公共服务 Public Services

"公共服务"展项轴测图
Axonometric drawing of the "Public Services"

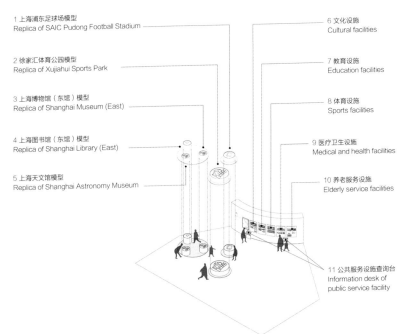

1 上海浦东足球场模型
Replica of SAIC Pudong Football Stadium

2 徐家汇体育公园模型
Replica of Xujiahui Sports Park

3 上海博物馆（东馆）模型
Replica of Shanghai Museum (East)

4 上海图书馆（东馆）模型
Replica of Shanghai Library (East)

5 上海天文馆模型
Replica of Shanghai Astronomy Museum

6 文化设施
Cultural facilities

7 教育设施
Education facilities

8 体育设施
Sports facilities

9 医疗卫生设施
Medical and health facilities

10 养老服务设施
Elderly service facilities

11 公共服务设施查询台
Information desk of
public service facility

展示内容

　　多层次、高水平的公共服务与社会保障体系是满足人民日益增长的美好生活需要的主要途径，公共服务品质越来越成为衡量人民群众获得感、幸福感、安全感的重要标尺。"公共服务"展项呈现上海在完善公平共享、弹性包容的基本公共服务体系，不断提升服务质量水平，增强市民获得感和满意度等工作中所做的努力。展项从五大公共服务设施——"文化设施""教育设施""体育设施""医疗卫生设施""养老服务设施"的发展与定位切入，结合典型范例，共同展示上海完善多层次高水平的公共服务和社会保障体系的发展目标、具体举措和实践成果，与上海作为社会主义现代化国际大都市的定位相适应。

模型集中展示具有代表性的文化设施及体育设施建设实例，包括上海图书馆（东馆）、上海天文馆、上海博物馆（东馆）、上汽浦东足球场、徐家汇体育公园等高等级公共服务设施。

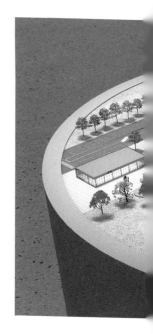

Content

A multi-layer, high-level public service and social security system can satisfy people's growing needs for a better life, and the quality of public services is becoming an important yardstick to measure people's sense of benefit, happiness and security. The "Public Services" exhibit area presents Shanghai's efforts to improve the basic public service system that is fair, shared, flexible and inclusive, its measures to continuously improve the quality and level of services, and to enhance people's sense of access and satisfaction. The area presents the five major public service facilities – "cultural facilities," "education facilities," "sports facilities," "medical and health facilities" and "elderly service facilities." Combined with examples, they show the development goals, concrete measures and practical achievements of Shanghai in improving the multi-level and high-level public service and social security system, and the positioning of Shanghai as a socialist modern international metropolis.

The replica focuses on premium cultural and sports facilities, including the Shanghai Library (East), Shanghai Astronomy Museum, Shanghai Museum (East), SAIC Pudong Football Stadium, Xujiahui Sports Park, among others.

上海图书馆（东馆）模型设计图
The replica of Shanghai Library (East)

上海天文馆模型设计图
The replica of Shanghai Astronomy Museum

上海博物馆（东馆）模型设计图
The replica of Shanghai Museum (East)

上海浦东足球场模型设计图
The replica of SAIC Pudong Football Stadium

展项解读

　　"公共服务"展项中的三个文化设施模型和两个体育设施模型，采取分散式布局。除徐家汇体育公园以 1:1200 比例呈现完整建筑群组外，其余四个案例均采用 1:400 比例的建筑单体模型。圆形元素应用于地贴和展台，大圆内部并置小圆，交错产生曲折蜿蜒的观赏路径，供观众进行全方位、多角度的深度观察。

Interpretation

The three replicas of cultural facilities and two replicas of sports facilities in the "Public Services" exhibit area adopt a decentralized layout. Except for the Xujiahui Sports Park, which is presented as a complete building group in 1:1200 scale, all the other four cases use 1:400 scale replicas of individual buildings. Circular elements are applied to the floor stickers and display stands, and small circles are juxtaposed inside the large circles to produce a winding viewing path for visitors to have an all-round and multi-angle in-depth observation.

徐家汇体育公园模型设计图
The replica of Xujiahui Sports Park

公众参与　Public Participation

"公众参与"展项轴测图
Axonometric drawing of the "Public Participation"

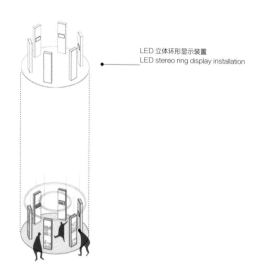

LED 立体环形显示装置
LED stereo ring display installation

12

展示内容

　　通过公众参与，城市规划才能够真正做到以人为本，进而促进城市文明的和谐健康发展。公众的关注、参与、讨论，有助于理解城市发展的宏观战略、微观操作，进而产生共同的认知与共情，让规划工作从实处出发，并落实于细微。"公众参与"展项依次展示"上海城市设计挑战赛""上海城市空间艺术季""Sea-Hi！论坛""品·筑""符号上海""漫行城市"等上海市规划和自然资源局打造的公众品牌项目成果，充分反映上海深厚的文化烙印和上海坚持"开门做规划"的工作原则。在完善自身建设的同时，促进各年龄段人群了解、支持和参与城市建设。

Content

Only through public participation can urban planning be truly human-centric and promote the harmonious and healthy development of urban culture. Public attention, participation and discussion will help the public understand the macro strategy and micro operation of urban development, and translate into common cognition. Only with the engagement of the public that the planning work can be carried out from the practical point of view and implemented in the smallest detail. The "Public Participation" area showcases the "Shanghai Urban Design Challenge," "Shanghai Urban Space Art Season," "Sea-Hi! Forum," " 'Pinzhu' Architecture Review," "Iconic Shanghai," "Roaming in Shanghai" and other public brand projects created by the Shanghai Municipal Bureau of Planning and Natural Resources. These efforts are not only the achievements of the Bureau but also reveal how the city engages people of all ages to understand and contribute to urban development.

展项解读

　　"公众参与"展项设置于二层展厅流线的终点。展项结合环形空间与开放式布局，通过双面 LED 屏幕打造出虚实交替的观展空间，生动展现城市规划工作中市民共同探索城市发展的实践成果。双面屏外侧为各项公众品牌项目主题海报的数字动态演绎。内侧则设置触摸屏展示项目的实景照片与影像资料等，观众通过观看和查询，能有效获取公众参与的渠道，并能加入到具体项目的实践中，与政府职能部门、专家、学者和相关从业者展开对话。

　　环形装置呼应二层展厅的圆形基础元素，六块宽 0.5 米、高 2.3 米的长方形立式展具环绕围合，供观众从各个角度进入穿行。内屏采用吸磁壁挂式设计，听筒与国际标准接轨较传统挂耳式耳机更具科技感与互动性，音质清晰而富有质感。

Interpretation

This area is tucked at the end of the flow line of the second-floor gallery, where the setting is an annular space with an open layout. The double-sided LED screen emanates a visual quality that mixes the virtual with the real, displaying the achievements of urban planning that involve public participation. The periphery of the screen rotates a raft of digital posters of public brand projects, while the touch screen in the middle presents their photos and videos. The visitors can find the channels of public participation through viewing and inquiring; when attending a particular project, they can talk to government functional departments, experts, scholars and professionals.

The circular installation echoes the circular elements on the second floor. Six rectangular vertical exhibition devices, each measuring 0.5-m by 2.3-m, surround and enclose each other, allowing the visitors to enter from various entries. The internal screen is a magnetic wall-mounted one, whose cutting-edge earphones feature an interactive nature unmatched by traditional hanging earphones. And their clear, rich acoustic quality can well challenge international standards.

外屏多媒体设计组图
Multimedia on outer screens

三层　创新之城

3F　An Innovative City

三层展厅平面图
Plan of the third-floor gallery

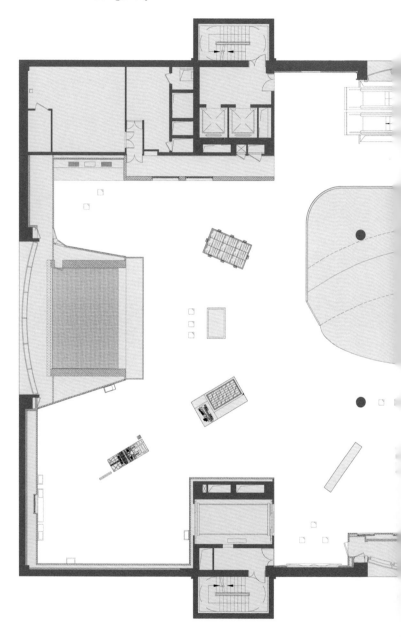

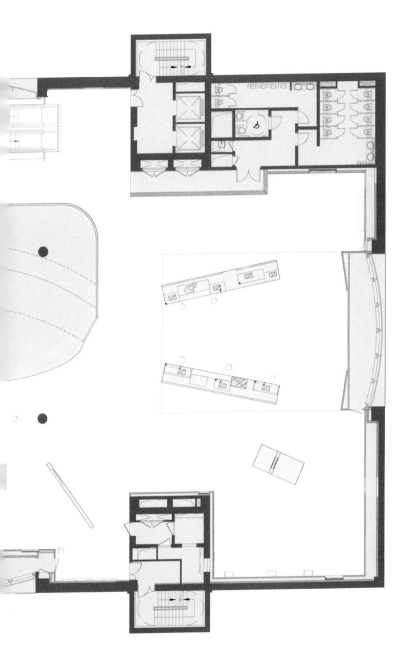

三层展厅轴测图
Axonometric drawing of the third-floor gallery

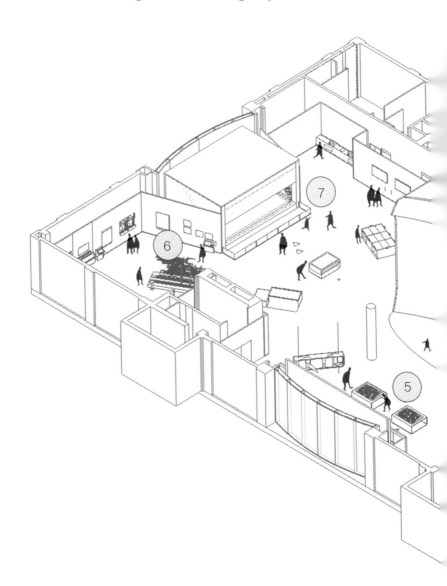

1 中国（上海）自由贸易试验区 China (Shanghai) Pilot Free Trade Zone
2 长三角一体化发展 Integrated Development of the Yangtze River Delta
3 科创中心 Science and Technology Innovation Center
4 服务提升 Service Improvement
5 五个新城 Five New Cities
6 国际枢纽 International Hub
7 城市数字化转型 Urban Digital Transformation

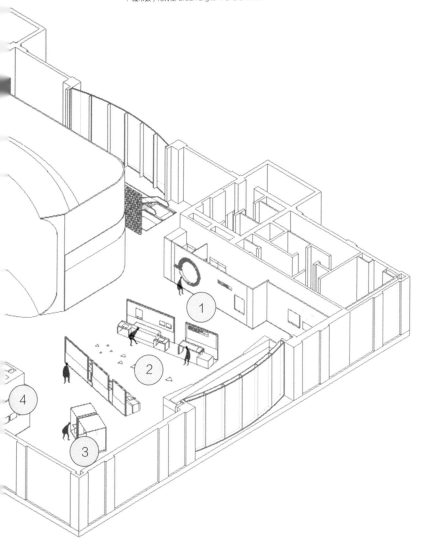

创新是引领发展的第一动力，是建设现代化经济体系的战略支撑。上海依托国家创新体系建设，将充分发挥服务长江经济带的龙头城市和"一带一路"建设桥头堡的作用，带动形成具有全球竞争力的长三角世界级城市群，基本建成具有全球影响力的科技创新中心。

三层展厅以"创新之城"为主题，充分体现上海作为"全国改革开放排头兵、创新发展先行者"，深入实施创新驱动发展战略。"创新之城"通过自贸试验区、长三角一体化发展、科创中心、服务提升、五个新城、国际枢纽、城市数字化转型七个展项，展现上海以"三大任务、一大平台"为战略引领，强化全球资源配置、科技创新策源、高端产业引领和开放枢纽门户四大功能，深化"五个中心"建设，加快建设具有世界影响力的国际数字之都，成为一座更具活力和更加繁荣的创新之城。

创新之城围绕着"创新"这个最核心、最本质的发展动力设计。在流线设计和展项布局方面，以三角形为基础元素，展台放射性设置，共同指向展厅空间的核心——数字沙盘。三角形自身具有稳定的几何美学特征，通过旋转角度，衍生出一定变化感与运动感。在视觉呈现方面，金色与灰色是三层展厅的主题色，深灰色墙面、深色天花板、灰色金属展台与金色三角形元素相结合，灰色地面作为基底，结合金属材质与创新性模型材料，营造现代简约、具有未来感的创新展示空间，彰显创新之城的活力与张力。

Innovation is the major development engine and a strategic support of a modern economic system. Relying on the nationwide innovation system, Shanghai gives full play to its leadership role of serving the Yangtze River Economic Belt and the powerhouse of "Belt and Road" construction, it has fueled the formation of a world-class city cluster in the Yangtze River Delta with global competitiveness, and has established itself as a scientific and technological innovation center with global influence.

The third-floor gallery offers an extensive survey of the "Innovative City," reflecting Shanghai's role as "the leader of reform and opening up and the pioneer of innovation and development" and the in-depth implementation of the innovation-driven development strategy. "An Innovative City" showcases seven exhibit areas, including "Pilot Free Trade Zone," "Integrated Development of Yangtze River Delta," "Science and Technology Innovation Center," "Service Improvement," "Five New Cities," "International Hub," and "Urban Digital Transformation," to demonstrate Shanghai's strategic leadership of "Three Major Tasks and One Major Platform," strengthening global with "three missions and one platform" as the strategic leadership, Shanghai will strengthen the four functions of global resource allocation, science and technology innovation source, high-end industry leader and open hub gateway, deepen the construction of "five centers," accelerate the construction of an international digital capital with world influence, and become a more dynamic and prosperous city of innovation.

This area is designed around the core element of innovation. In terms of flow design and exhibit layout, triangle is the basic element, and the radioactive layout of the display stand points to the core of the exhibition space - the digital sandtable. The triangle itself has stable geometric aesthetic characteristics, and through the rotation angle, a certain sense of change and movement is derived. In terms of visual presentation, gold and gray are the theme colors of the three-level exhibition hall. Dark gray walls, dark ceilings, gray metal display stands and golden triangle elements are combined with gray ground as the base, combined with metal materials and innovative replica materials to create a modern, simple and futuristic innovation display space, highlighting the vitality and tension of the city of innovation.

中国（上海）自由贸易试验区
China (Shanghai) Pilot Free Trade Zone

"中国（上海）自由贸易试验区"轴测图
Axonometric drawing of "China (Shanghai) Pilot Free Trade Zone"

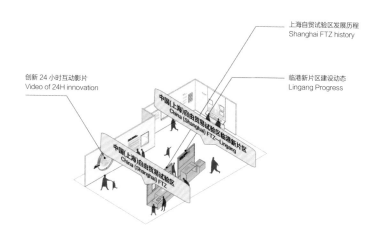

上海自贸试验区发展历程
Shanghai FTZ history

临港新片区建设动态
Lingang Progress

创新 24 小时互动影片
Video of 24H innovation

中国（上海）自由贸易试验区临港新片区
China (Shanghai) FTZ—Lingang

中国（上海）自由贸易试验区
China (Shanghai) FTZ

展示内容

　　建设自由贸易试验区是党中央、国务院在新形势下全面深化改革和扩大开放的一项重要战略举措。"中国（上海）自由贸易试验区"展项展示了中国（上海）自由贸易试验区的建设目标、建设历程、建设成果与制度创新，重点聚焦中国（上海）自由贸易试验区临港新片区的战略定位、规划布局和重点片区建设。展项集中呈现自上海自贸试验区挂牌成立以来的三个重要发展节点，并对区域规划及发展历程进行详细展示，呈现上海自贸试验区为促进中国经济全面融入世界经济体系而做出的不断探索。

　　临港新片区版块通过片区定位、发展历程、发展目标、先行启动区案例、建设动态等内容全面展示临港新片区在更深层次、更宽领域，以更大力度推进全方位高水平的对外开放，代表国家更好参与国际合作与竞争的重要作用与积极进展。

Content

The launch of the Pilot Free Trade Zone is an epic strategic move of the Central Government of China to double its efforts in reform and opening. This exhibit area displays the objectives, progress, achievements and institutional innovations of FTZ, with emphasis on its strategic positioning, planning and key area construction of its Lingang New Area. It gives nods to the three development milestones since the FTZ's inception, and displays in great detail the regional planning and development process. Visitors can find how the FTZ has scrambled to integrate the economy of China into that of the world.

Through the display of development history, goals, case study of experimental areas, and the updates, the exhibition of the Lingang New Area celebrates the Area's large role and progress in its monumental commitments to higher level opening to the rest of the world. It is indisputably a signature project of China in the realm of international cooperation and competition.

中国（上海）自由贸易试验区展项实景
"China (Shanghai) Pilot Free Trade Zone" exhibit area

临港新片区
Lingang New Area

展项解读

"中国（上海）自由贸易试验区"展项突出创新之城三角形的设计母题，L形展墙与一块立式展架自然围合出三角形观展空间，展架向中央数字沙盘聚拢，下部镶嵌深灰色钢化玻璃，玻璃自身的透明感增加了观展视线的穿透力，在弱化其本身的视觉重量的同时打通观众的视觉通廊。黑镜玻璃结合浅灰色基底展墙，深灰色金属展台，有机统一、相互协调。展项色彩沉稳大方，展示形式带有现代摩登感，为整层展厅奠定了创新与活力的基调。

在展项入口处设置了"创新24小时"的投影装置，聚焦人工智能、集成电路、生物医药三大创新产业。产业创新的场景跟随时钟的转动不断切换，体现了上海不断涌动的创新活力。

Interpretation

This exhibit area centers on the visual motif of triangle, where the L-shaped exhibition wall and a vertical exhibition stand naturally enclosing a triangular space. The transparency of the glass increases the permeability of the viewer's vision, lowering the visual weight while opening up the visual corridor. The black mirror glass combined with the light gray base wall and dark gray metal display stand, organic unity and mutual coordination. The colors of the exhibition items are calm and generous, and the display form has a modern sense of modernity, which sets the tone of innovation and vitality for the whole exhibition space.

Set up at the entrance, the projector features "24H innovation," focusing on the three innovative industries of artificial intelligence, integrated circuits and biomedicine. The scenes of industrial innovation follow the turning of the clock and keep switching, revealing the source of Shanghai's creative vitality.

长三角一体化发展
Integrated Development
of the Yangtze River Delta

"长三角一体化发展"轴测图
Axonometric drawing of
"Integrated Development of Yangtze River Delta"

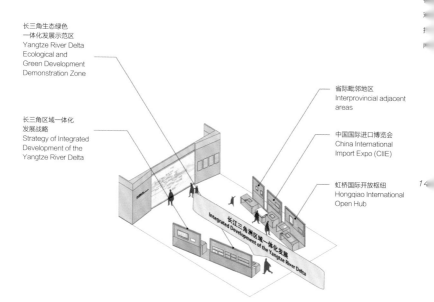

长三角生态绿色
一体化发展示范区
Yangtze River Delta
Ecological and
Green Development
Demonstration Zone

长三角区域一体化
发展战略
Strategy of Integrated
Development of the
Yangtze River Delta

省际毗邻地区
Interprovincial adjacent
areas

中国国际进口博览会
China International
Import Expo (CIIE)

虹桥国际开放枢纽
Hongqiao International
Open Hub

展示内容

　　长江三角洲地区（以下简称"长三角"）是我国
经济发展最活跃、开放程度最高、创新能力最强的区
域之一。2018 年 11 月，长江三角洲区域一体化发展
正式上升为国家战略，在国家现代化建设大局和全方
位开放格局中具有举足轻重的战略地位。"长三角一
体化发展"展示了长三角区域一体化发展的战略定位、
发展目标和规划布局，重点介绍长三角生态绿色一体
化发展示范区、虹桥国际开放枢纽、省际毗邻地区的
建设与实施举措。

43

长三角一体化发展展项实景
Exhibit area of Integrated Development of the Yangtze River Delta

Content

The Yangtze River Delta is one of the regions with utmost economic growth spurt, the highest degree of openness and the strongest innovation capacity in China. In November 2018, the integrated development of the Yangtze River Delta was declared a national strategy, giving the Delta a pivotal role in China's modernization drive and all-round opening campaign. The area showcases the strategic positioning, development goals and planning layout of the integrated development of the

一体化示范区生态格局示意图
Illustration of the ecological structure of the Integrated Development Demonstration Zone

Yangtze River Delta; emphasis is laid on the construction and implementation initiatives of the Yangtze River Delta Ecological and Green Development Demonstration Zone, the Hongqiao International Open Hub and border area between Shanghai and neighboring provinces.

展项解读

　　"长三角一体化发展"展项空间总体呈现三角形放射状格局,展项中央的巨幅屏幕展示长三角一体化发展的演绎影片。影片以"光与水的交响曲"为主题,展示长三角一体化的战略定位及发展愿景,将环境"颜值"转化为发展价值,营造古今辉映的江南发展新篇章。展项同时配有五台全频音箱及一台超低音音箱,打造出浑厚清晰的混响音效,结合开阔的中央观影空间,为观众提供了极佳观赏品质。

　　两排展台根据内容切分出多个长方形体块,并前后平移,以流动的造型打造由地面延伸到台面的连续形体,观众能够切身感受活力和动感的现代观展场景。展柜边缘作小型金色三角元素设计,三角形切面朝向数字沙盘,与整体轴线紧密呼应。

Interpretation

This area assumes a triangular, radial pattern. The giant screen at the center presents a video, "A Symphony of Light and Water," which depicts the integrated development of the Yangtze River Delta, with the theme of showing the strategic positioning. The theme of the film is "Symphony of Light and Water," showing the strategic positioning and development vision of the integration of the Yangtze River Delta, transforming environmental "value" into development value, and creating a new chapter of Jiangnan development reflecting the past and the present. The exhibition is also equipped with five full-range speakers and a subwoofer, creating a thick and clear reverberant sound, combined with an open central viewing space, providing visitors with excellent viewing experience. The two rows of display stands are divided into several rectangular blocks according to the content and panned back and forth to create a continuous form from the ground to the top of the stands in a flowing shape, allowing dynamic state-of-the-art scene. The corners of the display cases feature small golden triangular elements, and the latter face the digital sandtable, accentuating the overall axis.

"长三角一体化发展"影片
Video of the "Integrated Development of Yangtze River Delta"

科创中心
Science and Technology Innovation Center

"科创中心"轴侧图
Axonometric drawing of
"Science and Technology Innovation Center"

科创魔盒
Science and Technology Magic Box

"申城活力"投影
Video of "Shencheng Vitality"

展示内容

　　建设具有全球影响力的科技创新中心，是上海实施创新驱动发展战略的重要载体。"科创中心"展项围绕建设具有全球影响力的科技创新中心的总体目标，展示集成电路、生物医药、人工智能三大重点领域，着力培育科技创新策源功能，推进城市创新链与空间布局的耦合发展，并从科学创新、产业创新、技术创新、文化创意四个维度展开具体的创新举措与实践案例的陈述。

Content

Building a science and technology innovation center with global influence is an important mission of Shanghai to implement the innovation-driven development strategy. This area focuses on this overall goal and the three key areas of integrated circuits, biomedicine and artificial intelligence. It displays how Shanghai cultivates its function as a fountainhead of science and technology innovations, and how it promotes the development of urban innovation chain and spatial layout. The exhibits also showcase specific innovation initiatives and practical cases in four areas: scientific, industrial, technological innovation, and cultural creativity.

"科创中心"展项实景
"Science And Technology Innovation Center" exhibit area

展项解读

　　"科创中心"展项通过 L 形展墙与 M 形展台围合成布局集约的展示空间。展项中心区域的"科创魔盒"由灰色金属板包裹，两侧开敞，如同一个现代化驾驶舱。盒体由一个进深 1.28 米、高 2.05 米的小盒体及一个进深 1.76 米、高 2.05 米的大盒体连接而成，两个盒体通过 55 英寸显示屏生动演绎《人工智能》和《上海科创人》影片。《上海科创人》以真实人物访谈的方式，还原园区内科创工作者的真实工作场景，帮助观众走进充满活力的科创生活。外侧展墙的三块投影屏演绎《申城活力》动态影像，影像采用艺术化与科技化的表达形式，将不同人物迈步的剪影与上海各个创新承载区进行交叠展现。

Interpretation

The "Science and Technology Innovation Center" exhibit area is a compact space enclosed by an L-shaped exhibition wall and an M-shaped display stand. The "Science and Technology Magic Box" in the center of the exhibit area is wrapped by gray metal plates and open on both sides in the manner of a modern cockpit. The box consists of a small box with a depth of 1.28 meters and a height of 2.05 meters and a large box with a depth of 1.76 meters and a height of 2.05 meters, which are connected by a 55-inch (i. e. diagonal length of 1.4 meters) display that vividly portrays the "*Artificial Intelligence*" and "*Shanghai Science and Technology Creators*." The "*Shanghai Science and Technology Creators*" film uses interviews to restore the real working scenes of science and technology creators in the park, helping visitors to enter the vibrant life of science and technology creation. The three projection screens on the outer exhibition wall interpret the dynamic image of "*Shencheng Vitality*," which adopts artistic and technological expressions to show the overlapping silhouettes of different characters moving forward and various innovation-bearing zones in Shanghai.

服务提升　Service Improvement

"服务提升"轴侧图
Axonometric drawing of "Service Improvement"

展示内容

　　上海加快国际金融中心建设、打造高品质商务集聚区以及建设世界著名旅游目的地。"服务提升"展项从金融商务、旅游休闲两个方面展开。金融商务版块，围绕加快国际金融中心建设目标，聚焦空间创新和制度创新亮点。旅游休闲版块，围绕建设世界著名旅游目的地城市的目标，聚焦上海国际旅游度假区、佘山国家旅游度假区等重点区域的建设实践。

Content

Shanghai is accelerating its effort to establish itself as an international financial center, home to high-profile business clusters and a tourist destination of world renown. This area is launched from two dimensions: financial business, tourism and leisure. The financial business section focuses on spatial and institutional innovation around the goal of accelerating the construction of international financial center. The tourism and leisure section focuses on the construction practice of Shanghai International Tourism Resort, Sheshan National Tourism Resort and other key areas around the goal of building a world-renowned tourism destination.

展项解读

　　"服务提升"展项由一面 L 形展墙构成，展示空间上自然转折，展示内容则以简洁明了的图文形式清晰展示上海在金融商务和服务休闲方面的创新亮点和特色案例。电子触摸屏围绕旅游休闲的典型案例展开，让观众深入了解上海国际旅游度假区核心区的重点项目分布、佘山国家旅游度假区的项目概况，以及南京东路东拓项目的特色与亮点。

Interpretation

This area includes an L-shaped wall with a natural twist in the display space, while the content of the display highlights the innovative and characteristic cases of Shanghai's financial business and service-leisure sectors in a clear, concise format. The touch screen focuses on typical cases of travel and leisure, e.g., the key projects in the core area of the Shanghai International Resort, the Sheshan National Tourism Resort, and the extension project of East Nanjing Road.

服务提升展项设计图
The design of Service Improvement

五个新城　Five New Cities

"五个新城"轴侧图
Axonometric drawing of "Five New Cities"

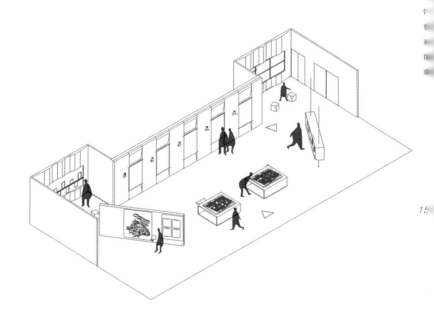

展示内容

　　"五个新城"的建设是上海承担国家战略、服务构建新发展格局的重要载体，是上海面向未来的重大战略选择，是上海服务辐射长三角的战略支撑点。"五个新城"展项从国家战略展开，系统呈现上海"主城区—新城—新市镇—乡村"的市域城乡体系建设以及新城发展规划，重点展示上海在"十四五"期间加快形成"中心辐射、两翼齐飞、新城发力、南北转型"的空间新格局，同时聚焦五个新城发展目标、建设动态和重大项目。

Content

The construction of five new cities is an important effort of Shanghai to undertake national strategy and to serve to a new development pattern. It is also a major strategic choice of Shanghai to face the future, and a strategic support point to serve the Yangtze River Delta. This area starts with the national strategy and presents the "main city - new cities - new towns - village" construction system of Shanghai. It focuses on Shanghai's new spatial pattern with a radiating center, two burgeoning wings, powerful new cities and south-north transformation during the 2021-25 period, and the development goals, construction dynamics and major projects of the five new cities.

"五个新城"展项实景
"Five New Cities" exhibit area

嘉定新城总体城市设计土地使用规划示意图
Urban design and land use planning of Jiading New City

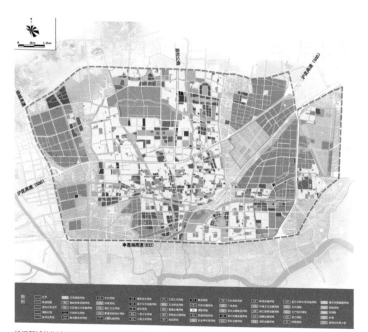

松江新城总体城市设计土地使用规划示意图
Urban design and land use planning of Songjiang New City

青浦新城总体城市设计土地使用规划示意图
Urban design and land use planning of Qingpu New City

奉贤新城总体城市设计土地使用规划示意图
Urban design and land use planning of Fengxian New City

南汇新城总体城市设计土地使用规划示意图
Urban design and land use planning of Nanhui New City

展项解读

  "五个新城"展项由三面展墙、五个模型展台、一个立式展板和一个悬挂式展板有机组成，若干方形坐凳点缀其间，为观众提供充足的参观与停留空间。展项设计继续延用三角形母题，模型展台、立式展板、悬挂式展板和方形坐凳均设置金色三角形包边，其切面均朝向中央数字沙盘，更加突出了轴向放射性的空间布局特色。案例模型采用 3D 树脂材料打印，与金属展台和谐统一，体现出科技感与未来感。悬挂式展板通过透明亚克力玻璃印刻相关图文内容，整体通透，氛围明快。

Interpretation

This area provides sufficient space for visitors to see or sit: Three walls, five replica tables, one vertical display board and one hanging display board are set up with square stools dotted among them. The design still embraces the triangle motif: the display stands, vertical and hanging display board and square sitting stool are all decorated with golden triangular edges, and their cut surfaces face the digital sandtable in the middle, highlighting the layout of axial radioactivity. The 3D-printed resin replica echoes with the metal table, reflecting a futuristic sense of technology. The hanging board is a transparent acrylic panel with images and text, looks lucid and lively.

奉贤望园森林芯中央活动区
Wangyuan Forrest Core in Fengxian New City Central Activities Zone

南汇世界顶尖科学家社区产业社区
World Laureates Community in Nanhui New City Industrial Community

松江云间站城核中央活动区
Yunjian Station-City Integration in Songjiang New City Central Activities Zone

嘉定远香文化源中央活动区
Yuanxiang Cultural Source in Jiading New City Central Activities Zone

模型仅为概念设计方案
The models are conceptual design.

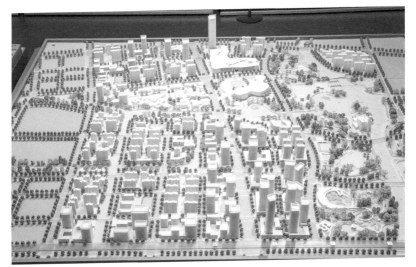

青浦上达创芯岛中央活动区
Shangda Chuangxin Island in Qingpu New City Central Activities Zone

中段展墙纵向设置五块 86 英寸触摸屏，正向均等排布，全方位展示每个新城的规划布局和建设动态，可同时满足多人查询需求。考虑到各年龄层次观众操作的舒适性，触点位置集中设置于下半部分，便于特殊人群和儿童查询。西侧展墙设有六块 55 英寸屏幕组合的超大型 LED 触摸屏，应用红外感应技术，展示五个新城总体设计和重点地区城市设计方案。

Five 86-inch (2.18 meters diagonally) touch screens are installed in the middle section of the exhibition wall. They are evenly distributed to brief on the planning and construction updates of each new city, ready for the enquiries of several visitors at the same time. Considering facilitating visitors of all ages—especially children—the buttons are set in the lower half of the screens. The west of the exhibition wall is equipped with a mammoth LED touch screens made up of 55-inch screens; infrared induction technology is employed to display the overall design of the five new cities and urban design plans of key areas.

五个新城触摸屏展项实景
Five New Cities touch screens exhibit area

国际枢纽
International Hub

"国际枢纽"轴侧图
Axonometric drawing of "International Hub"

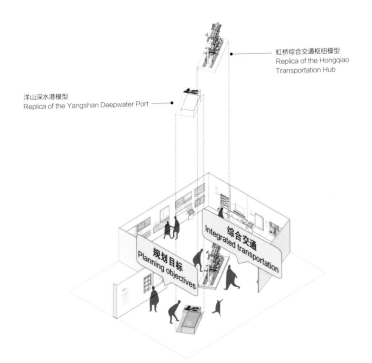

虹桥综合交通枢纽模型
Replica of the Hongqiao
Transportation Hub

洋山深水港模型
Replica of the Yangshan Deepwater Port

综合交通
Integrated transportation

规划目标
Planning objectives

展示内容

　　上海强化开放枢纽门户功能要求。"国际枢纽"展项重点展示上海建设更开放的国际枢纽规划目标和综合交通体系建设。规划目标从强化亚太地区航空门户地位、推动国际海港枢纽功能升级、增强铁路枢纽辐射服务能力、提升信息通信枢纽服务水平四个方面进行解读。综合交通体系建设主要涵盖实施公交优先战略、优化道路交通功能、发展现代货运系统、积极适应新兴技术发展、公共交通枢纽综合开发五方面内容。

Content

To echo with Shanghai's commitment to enhancing its function as a hub and gateway for the opening of China, this area introduces the planning objectives and construction of a comprehensive transportation system.

国际枢纽展项实景
"International Hub" exhibit area

The planning objectives include ambition in four aspects: as an aviation gateway in the Asia-Pacific region, an international seaport hub, a railroad hub that radiates service capacity, and an information and communication hub. The construction of the integrated transportation system covers five aspects: giving priority to public transport, optimizing road traffic, developing a modern freight system, embracing the development of emerging technologies, and the integrated development of public transportation hubs.

展项解读

　　"国际枢纽"包括五面展墙、两个实体模型、一个投影区以及一个墙体模型，视觉设计上延续了三角形元素。展区入口处的洋山深水港模型和虹桥综合交通枢纽模型均布局在放射形轴线上，营造了动态、活力的展示空间。投影区动态展示了上海航空、航运辐射网络，直观展现上海开放国际枢纽的互联互通。墙体模型展示了城市街道设计的优化，使观众能够近距离地感受街道空间。

Interpretation

This area includes five exhibition walls, two replicas, one projection area and a wall-based replica. T Triangles are the dominant elements in the design. The replicas of the Yangshan Deepwater Port and of the Hongqiao Transportation Hub at the entrance of the area are laid out on a radial axis, creating a dynamic and energetic feel. The projection area alternately shows the city's aviation and shipping radiation network, demonstrating Shanghai's connections as an open international hub. The wall-based model can tell you how the city optimizes its design of street network.

虹桥综合交通枢纽模型设计图
The replica of the Hongqiao Transportation Hub

技术特色

　　虹桥综合交通枢纽模型将现代化科技与艺术化表达进行有机结合，采用叠置方式，通过光敏树脂三维打印、树脂浇筑水面、亚克力雕板等技法进行制作。为增强观众的场景体验，模型将数据信息进行分层、分区呈现，附加至各层亚克力切片上，同时通过透明 LED 大屏幕及 AR 增强显示的技术手段，实现对虹桥综合交通枢纽自身智慧系统以及建造逻辑的多手段、多层次、全方位的解读与呈现。

Technical Features

The replica of the Hongqiao Transportation Hub organically combines modern technology with artistic design. Produced into several layers to represent different floors, it uses such technologies as photosensitive resin 3D printing, resin-poured water surface, and acrylic plate carving. To make the visual effect more vivid, the replica presents data information on different layers and partitions before attaching it to each layer of acrylic pieces. The transparent LED screen and AR technology help demonstrate the Hongqiao Transportation Hub's own intelligent system and construction logic with multiple means.

城市数字化转型
Urban Digital Transformation

"城市数字化转型"展项轴侧图
Axonometric drawing of "Urban Digital Transformation"

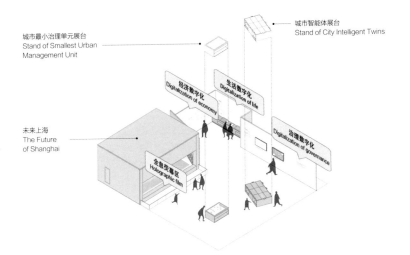

城市最小治理单元展台
Stand of Smallest Urban
Management Unit

城市智能体展台
Stand of City Intelligent Twins

经济数字化
Digitalization of economy

生活数字化
Digitalization of life

治理数字化
Digitalization of governance

未来上海
The Future
of Shanghai

全息荧幕区
Holographic film

展示内容

上海围绕全面推进城市数字化转型的重大战略决策，聚焦经济、生活、治理三大领域，加快建设具有世界影响力的国际数字之都。"城市数字化转型"展项围绕"经济数字化""生活数字化""治理数字化"三方面展示数字化转型建设的总体框架、目标愿景和行动方案，同时选取"城市智能体""城市最小管理单元（南京大楼）"等典型案例体现数字化在城市中的应用。《未来上海》主题影片，诠释在数字化转型战略的引导下，对上海未来城市美好生活的畅想。

Content

Shanghai focuses on three major areas: economy, life and governance, and accelerates the construction of an international digital capital with world influence, based on the major strategic decision to comprehensively promote the digital transformation of the city. This exhibit area focuses on "Digitalization of Economy," "Digitalization of Life" and "Digitalization of Governance." The exhibition showcases the overall framework, target vision and action plan of digital transformation, while typical cases such as 'City Intelligent Twins" and "Smallest Urban Management Unit (Nanjing Building)" are selected to reflect the application of digitalization in the city. The theme film "*The Future of Shanghai*" will explain the future of Shanghai's city life under the guidance of digital transformation strategy.

展项解读

　　"城市数字化转型"展项引入多元展示方式，包括墙面交互展示区的"经济数字化""生活数字化""治理数字化"，案例互动展台的"城市智能体""城市最小管理单位"，以及全息放映区的《未来上海》，整体秩序清晰，空间丰富多变。墙面互动展示区运用触控感应、卡牌感应技术，增强观众的互动体验。《未来上海》影片采用墙屏、地屏和全息投影三屏联动的视觉呈现方式，营造未来感和科技感。"城市最小管理单元"案例互动展台真实还原南京大楼内部场景，引导观众直观感知城市数字化治理的最新探索与创新实践。

Interpretation

This exhibit area employs multiple display means, including wall-hung interactive devices—showing the "Digitalization of Economy," "Digitalization of Life' and "Digitalization of Governance," the interactive display stands—devoted to "City Intelligent Twins" and "Smallest Urban Management Unit," and the screen area, which airs the holographic film of "The Future of Shanghai." All these items are arranged

城市数字化转型展项实景
"Urban Digital Transformation" exhibit area

in a clear overall order, giving a rich, varied feeling of space. The interactive display wall uses touch- and card-sensitive technology to enable you to interact with it. The "Future Shanghai" film is presented on three screens, on the wall, the floor and through holographic projection, creating a sense of future and technology. The interactive display stand of the "smallest urban management unit" case truly restores the interiors of the Nanjing Building, so you can intuitively feel the latest exploration and innovative practices of digital urban governance.

技术特色

　　"城市智能体"案例互动展台选用水晶玻璃材质搭建，通过九块 86 英寸大型透明触摸屏拼接，结合 AR 交互技术展现未来感和科技感。

　　《未来上海》全息荧幕长 8.8 米，宽 0.6 米，高 4.4 米，采用墙屏、地屏和全息投影三种不同的影像介质描摹了一个更具活力、更富温度和更可持续的未来之城。该区域融合了三维计算机影像及全息荧幕技术，影像通过顶面屏幕反射至斜拉的透明成像膜上，形成裸眼三维的视觉效果，并在此基础上，增加一道幻影成像屏幕的画面层次，使画面内容可以在多维空间中相互切换融合，丰富了整体视觉语言且具有形式感，营造了新的视觉体验。

"城市智能体"互动展台细节
Details of The "City Intelligent Twins" interactive display stand

Technical Features

The "City Intelligent Twins" interactive display stand, built with crystal glass, is joined by nine 86-inch transparent touch screens and taps into AR interactive technology.

"The Future of Shanghai" holographic screen, 8.8-m by 0.6-m by 4.4-m, uses three different media: wall screen, floor screen and holographic projection to depict a dynamic, sustainable and resident-friendly city of the future. This area combines 3D computer graphic and holographic screen technology: The images are reflected to the diagonal transparent imaging film through the top screen, forming a naked eye 3D visual effect, and on this basis, adding a phantom imaging screen level, so that the content of the screen can be switched and integrated in multi-dimensional space, which enriches the overall visual language and has a sense of form, creating a brand-new visual experience like no other.

黄浦江
Huangpu River

《未来上海》全息荧幕
"The Future of Shanghai" holographic screen

南京大楼多媒体展台实景
Multimedia display stand of Nanjing Building

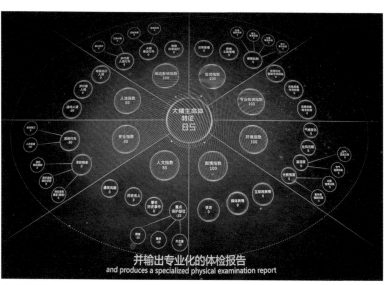

南京大楼多媒体演绎
Multimedia display of Nanjing Building

四层 生态之城

4F   An Ecological City

四层展厅平面图
Plan of the fourth-floor gallery

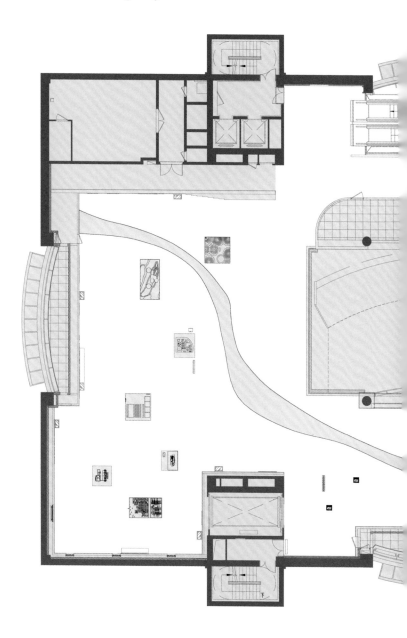

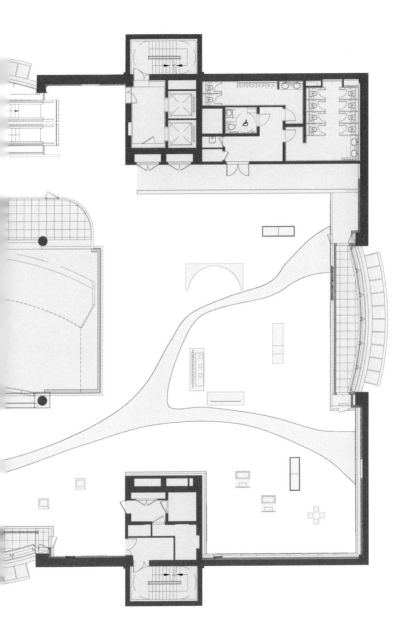

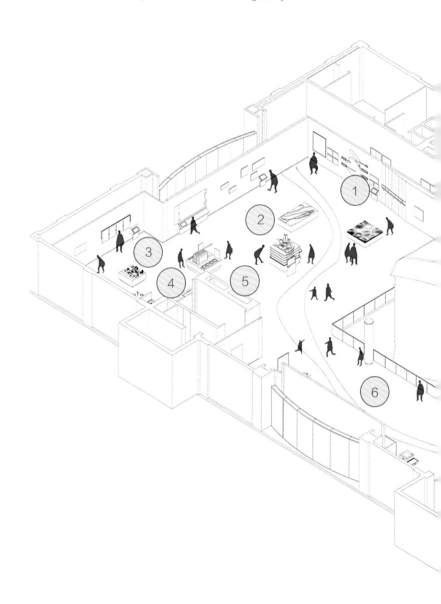

1 岁月的积淀 Precipitation of Ages
2 广域的空间 Extensive Space
3 底线约束 Bottom-line Control
4 绿色低碳循环 Green, Low-carbon Circulation
5 城市安全 Urban Security
6 生态空间规划 Ecological Space Planning
7 城乡公园体系 Urban-Rural Park System
8 乡村振兴 Rural Revitalization
9 崇明世界级生态岛 Chongming World-class Ecological Island
10 上海"2035"：规划实施框架体系 "Shanghai 2035"-The
Framework System of Implementation

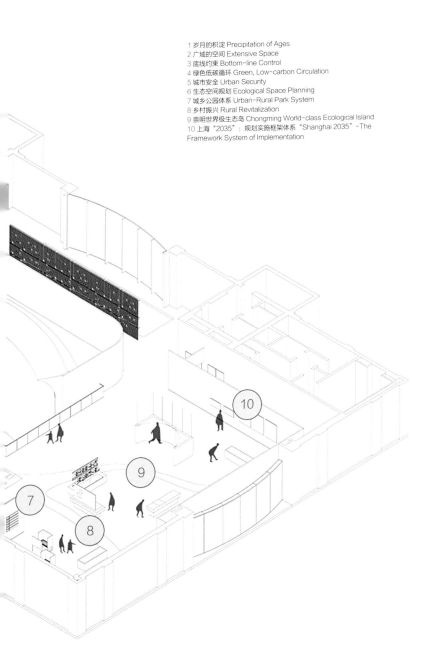

生态之城的目标是"尊重自然、顺应自然、保护自然，探索人与自然和谐共生之路"。上海坚持节约资源和保护环境的基本国策，坚持节约优先、保护优先、自然恢复为主的方针，坚持紧约束下的睿智发展，构造城市生态安全屏障，不断提升城市适应能力和韧性，成为引领国际超大城市绿色、低碳、安全、可持续发展的标杆。生态之城由"岁月的积淀"的宏观生态基底开启，引出"广域的空间、底线约束、绿色低碳循环、城市安全、生态空间规划、城乡公园体系、乡村振兴、崇明世界级生态岛"等展项，为观众呈现上海构筑多层次、成网络、功能复合的生态空间体系，成为一座更具韧性和更可持续的生态之城的美好愿景。同时四层展厅还设置了"上海2035：规划实施框架体系"和结语，作为整个展陈的总结。

　　方形为四层的基础元素，结合灰绿色的主色调，表达出对于生态、自然的美好向往。水是激活生态基底的主动脉，展台错落有致的布局模拟依山傍水、梯田乡宅的自然风貌，并辅以大自然的色调和环保材质，营造清新明快、绿意盎然的生态展示空间，呈现上海傍水而生的悠久历史，回应绿色生态的发展理念。

　　生态之城展厅基于交互式理念进行智慧展陈设计，通过展厅贯穿地面空间流动的"河流"，自然地串联起大小错落的展台。"河流"通体由树脂浇筑而成，采取精致工艺按"基层—保护层—边缘层"层层铺筑，一体成形。同时"河流"元素结合32台投影仪使其产生自然涌动感，投影影像设计来源于山、水、林、田、湖、草等自然生态要素之间相互联系、相互制约所产生的平衡感。互动投影区通过红外线传感器实时捕捉参与者，利用投影仪和摄像机进行捕捉拍摄，投影仪主要负责拍摄内容，摄像机则负责捕捉参与者的活动轨迹。在构成"地面—展厅空间—展墙"的完整秩序空间的同时，为展厅空间增添了生机与活力。

"河流"元素结合32台投影仪产生出自然涌动感
A "natural" river from design elements and 32 projectors

An ecological city aims to "respect nature, follow the rules of nature, protect nature, and explore the way for harmonious coexistence between man and nature." Shanghai adheres to the basic national policy of resource conservation and environmental protection, and the guideline of prioritizing

saving, protection and natural recovery. Encouraging sustainable development under restriction, it is building an ecological barrier for urban security, continuously improving its adaptability and resilience, and become a green, low-carbon, safe and sustainable benchmark of leading international metropolis. The section "An Ecological City" starts with the macro ecological base "Precipitation of Ages," which leads to units "Extensive Space" "Bottom-line Control" "Green, Low-carbon Circulation" "Urban Security" "Ecological Space Planning" "Urban-Rural Park System" "Rural Revitalization" and "Chongming World-class Ecological Island." It presents Shanghai's vision of building a multi-layered, networked and functional ecological space system and a more resilient and sustainable ecological city. There is also a unit called "Shanghai 2035"-The Framework System of Implementation" on the fourth floor, with a conclusion of the whole exhibition.

With square as the basic element and celadon as the dominant color, the fourth floor embodies the beautiful yearning for ecology and nature. Water is the key to activating the ecological base. The well-arranged layout of the stand simulates the scene composed of mountains, rivers, townhouses and the terraced fields. Supplemented with the colors of nature and environmentally-friendly materials, it creates a fresh and lively green exhibition space, echoing the long history of Shanghai thriving near the water as well as the philosophy of green development.

The "Ecological City" features a smart design based on interaction. A "river" flowing across the hall naturally connects the stands of different sizes. Its integrated body includes three layers, namely, the base layer, the protective layer and the edge layer, made by casting resin through exquisite workmanship. Besides, there are 32 projectors working to create a natural sense of surging, inspired by the sense of balance generated by the mutual connection and restriction among natural elements, such as mountains, water, forests, fields, lakes, and grasses. The interactive projection area captures participants in real time through infrared sensors, with projectors and cameras for shooting. The former is mainly responsible for shooting, while the latter captures the activity trajectory of the participants. While forming an orderly space of "ground, exhibition space and exhibition wall," it adds vitality to the whole space.

岁月的积淀　Precipitation of Ages

"岁月的积淀"轴测图
Axonometric drawing of "Precipitation of Ages"

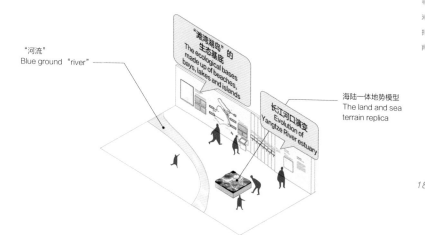

"河流"
Blue ground "river"

"滩湾湖岛"的
生态基底
The ecological bases
made up of beaches,
bays, lakes and islands

长江河口演变
Evolution of
Yangtze River
estuary

海陆一体地势模型
The land and sea
terrain replica

展示内容

时光沉淀，海浪冲刷，江海孕育形成上海今日独特的自然条件和生态基底。"岁月的积淀"展项基于时间与空间两个维度，依循海陆变迁、地势演变、上海成陆过程的叙述顺序，介绍上海在自然演变下构筑的"滩湾湖岛"生态基底。

处于长江入海口的上海具有江海交汇的特殊区位特征，以及独特的自然条件和生态基底。展项主要呈现距今二百万年来七次海侵及海退的过程、距今两万年来长江河口经历的地势演变和上海的成陆过程，反映上海随着长江河口江海岸线的演变，逐步构筑了"滩、湾、湖、岛"的生态基底，孕育了"山、水、林、田、湖、草"的生态要素，形成"农林水复合、林田湖相间"的生态特色。

Content

Years of ebb and flow has seen rivers and seas give birth to the unique natural conditions and ecological base of Shanghai today. Spanning time and space, this area introduces the ecological bases made up of beaches, bays, lakes and islands developing over time in the narrative sequence of sea and land changes, terrain evolution, and Shanghai's land-forming process.

Shanghai, which is located at the mouth of the Yangtze River, enjoys the unique intersection of rivers and seas, natural conditions and ecological bases. This area mainly presents the process of seven transgressions and regressions in the past two million years, and the topographical evolution of the Yangtze River Estuary and the land-forming process of Shanghai in the past 20,000 years. It reflects the gradual formation of the ecological bases made up of beaches, bays, lakes and islands along with the evolution of the coastline of the Yangtze River Estuary. This has also been bred such ecological elements as mountains, water, forests, fields, lakes and grass, leading to the ecological feature of "combination of agriculture, forestry and water, with forest, field and lake connecting each other."

展项解读

　　"岁月的积淀"展项空间布局简洁，展墙版面、图像排布、展板形式以方形为主元素，以变化多元的展墙、凹凸起伏的展板、可视化模型展台的表达方式丰富了空间展示层次，科普与艺术的融合能够给观众带来更具美感的观展体验。

　　海陆一体地势模型的演绎通过地面高差的艺术化处理，等高线围合出的旋涡式形状，生动模拟出海水侵刷后的自然样貌，使地势变化活灵活现。同时设有多媒体互动装置，转动按钮就能直观感受各时代的海平面变化，在上海成陆过程这一广阔的时空中自由穿梭。

岁月的积淀展项实景
"Precipitation of Ages" exhibit area

Interpretation

The spatial layout is simple. The wall, the images, and the exhibition boards are dominated by square elements. The diverse exhibition walls, the wavy exhibition boards, and the stands with visual replicas enrich the spatial display. The integration of art and popular science brings a more aesthetic viewing experience to the public.

The land and sea terrain replica presents vivid terrain changes by simulating the natural appearance after brushing by the sea, by virtue of the artistic treatment of the ground height difference, and the vortex shape created by the contour lines. Meanwhile, there is a multimedia interactive device, which allows visitors to intuitively feel the changes of sea level in various eras by turning the button, enjoying the process of Shanghai becoming a land.

海陆一体地势模型多媒体演绎
Multimedia of land and sea terrain replica

广域的空间　Extensive Space

"广域的空间"轴测图
Axonometric drawing of "Extensive Space"

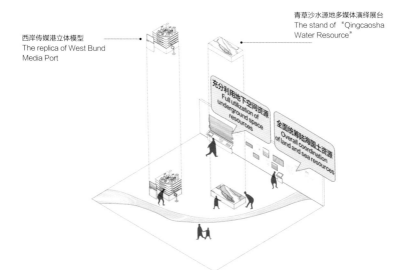

西岸传媒港立体模型
The replica of West Bund
Media Port

青草沙水源地多媒体演绎展台
The stand of "Qingcaosha
Water Resource"

充分利用地下空间资源
Full utilization of
underground space
resources

全面统筹陆海国土资源
Overall coordination
of land and sea resources

展示内容

　　上海积极实施陆海统筹发展战略，构建陆海开放型国土开发格局。"广域的空间"围绕上海建设陆海开放型国土开发格局的目标，全面展示上海充分利用地下空间资源，积极推进城市地上、地下空间一体化开发利用的主要工作情况。重点聚焦上海对于陆海国土资源全面统筹及地下空间资源的综合开发与分层利用。

Content

Shanghai implements the land-sea coordinated development strategy and aims to build an open land-sea development pattern. "Extensive Space" focuses on such a goal, and comprehensively demonstrates how Shanghai makes full use of underground space resources and actively promotes the integrated development and utilization of above-ground and underground urban spaces. The highlight is the overall planning of land and sea resources and the comprehensive development and hierarchical utilization of underground space resources in Shanghai.

广域的空间展项实景
"Extensive Space" exhibit area

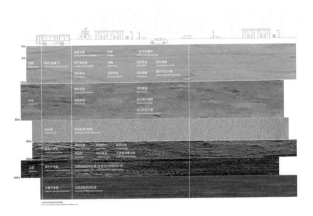

上海市地下空间分层布局示意图
Illustration of multilevel underground space of Shanghai

展项解读

　　"广域的空间"展项重点聚焦模型展台的设计，塑造出灵动的观展效果。

　　"青草沙水源地展台"通过多媒体演绎，真实模拟自然空间与形态，讲述水库应用潮汐动力的传输过程。同时，影像通过动画演绎方式展现水生物的活动场景，体现生物多样性。

　　"西岸传媒港模型"采用分层切片的处理方式，使地下三层空间及地面一层空间功能、结构关系及使用现状一目了然。模型选用磨砂亚克力材质表达建筑结构，彩色亚克力材质表达交通流线。模型展台采用可升降装置，在徐徐上升或下降的过程中，叠加出不同角度的生动立体效果，极具科技感。为了让观众直观了解每一层空间功能，展台同时配置专用平板电脑，将 AR 演绎结合 55 英寸立式透明屏，清晰呈现传媒港复杂的地面地下综合开发信息。

Interpretation

The focus is the replica that creates a flexible viewing effect.

Supported by multimedia, the stand "Qingcaosha Water Resource" simulates the natural space and form, and describes the transmission through tidal power in the reservoir. Meanwhile, the activity of aquatic creatures is demonstrated through animation, reflecting biodiversity.

The replica of West Bund Media Port features four layers, namely, three floors underground and one above the ground, with clear functions, structural relationships and usage. Frosted acrylic material is used to present the building structure while the colored acrylic material for the traffic flow. There is a lifting device to show the buildings from different angles through slowly rising or falling, which is very technological. For an intuitive view of the spatial functions of each floor, the stand is also equipped with a dedicated tablet computer, which combines AR with a 55-inch vertical transparent screen to demonstrate the complex, comprehensive development above ground and underground of the media port.

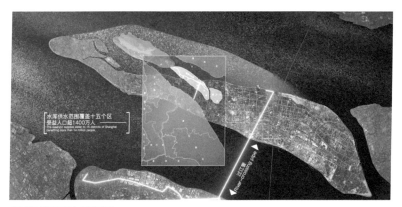

"青草沙水源地展台"多媒体演绎
"Qingcaosha Water Resource" shown in multimedia

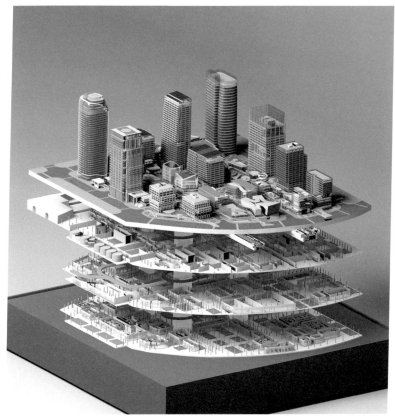

通过 iPad 结合 AR 演绎西岸传媒港地下空间利用
The underground space utilization of West Bund Media Port shown in iPad and AR device

底线约束
Bottom-line Control

"底线约束"轴测图
Axonometric drawing of "Bottom-line Control"

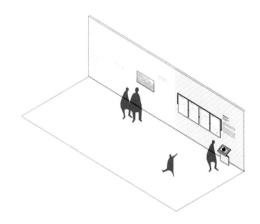

展示内容

　　突出底线约束，把保护城市生态环境和保障城市安全放在优先位置。"底线约束"展项重点展示上海人口规模、土地资源、生态环境和城市安全四条安全底线，坚持走低碳绿色有韧性的可持续发展道路，推进人与自然和谐共生。

Content

Bottom-line Control is highlighted, giving priority to the protection of the urban environment and the guarantee of urban security. This area focuses on the four bottom lines of Shanghai, namely, population, land, environment and security. Shanghai is adhering to a low-carbon, environmentally-friendly and resilient approach to sustainable development in order to promote the harmonious coexistence of man and nature.

"底线约束"展项实景
Exhibition wall of "Bottom-line Control"

展项解读

　　"底线约束"采用展墙与电子媒介的结合方式丰富展示内容。触摸屏详尽解读相关政策，有效拓展了展示空间与展陈深度。为进一步突出底线约束的重要性，展项的墙面背景采用深绿色，与相邻展项的浅绿色背景形成对比，丰富了观展视角效果。

Interpretation

The exhibition walls work with the electronic media to enrich the display. The touch screen offers detailed interpretation of relevant policies, effectively expanding the space of display and the depth of exhibition. To highlight the importance of the bottom-line control, the wall in dark green forms a contrast with the light green background of the adjacent unit to enrich the perspective.

"底线约束"绿色展墙与相邻展项形成对比关系
Contrastive colors of the exhibition walls of different exhibit areas

绿色低碳循环
Green, Low-carbon Circulation

"绿色低碳循环"轴测图
Axonometric drawing of "Green, Low-carbon Circulation"

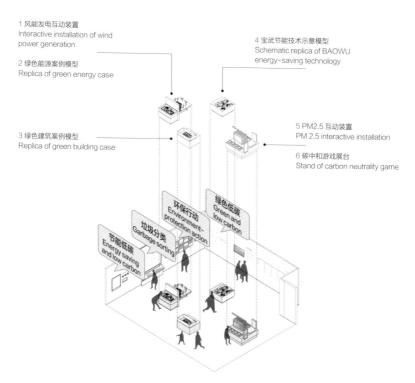

19

1 风能发电互动装置
Interactive installation of wind power generation

2 绿色能源案例模型
Replica of green energy case

3 绿色建筑案例模型
Replica of green building case

4 宝武节能技术示意模型
Schematic replica of BAOWU energy-saving technology

5 PM2.5 互动装置
PM 2.5 interactive installation

6 碳中和游戏展台
Stand of carbon neutrality game

展示内容

"绿色低碳循环"展项从生态发展指标入手，分别围绕绿色、低碳、循环三方面展示上海的探索与实践。

绿色低碳循环多媒体基于双碳内容，从科普维度展示大量实际案例，深入浅出地解读"碳达峰碳中和"的概念，展现上海未来绿色低碳循环发展的美好蓝图。

"环保行动"展项聚焦自 2000 年来，上海滚动实施的环保行动计划。通过社会共治、和谐共生、系统治污、绿色赋能四大重要措施的解读，展示上海在解决环境问题和加强城市环境治理方面的不懈努力，呈现环保行动实施的成效。

"垃圾分类"展项展示上海在"减量化、资源化、无害化"原则下建立完善的垃圾分类系统，构建生活垃圾全程分类体系。通过四张纵向全程分类流程图，对四类垃圾的循环利用做出直观表达。

"节能低碳"展项聚焦绿色低碳生产生活方式。展项通过数据图表清晰呈现出浅层地热资源在建筑应用、室外换热系统、地源热泵项目等方面的利用状态，并完整展示上海地区浅层地热能开发利用项目分布现状。

Content

Based on the indicators of ecological development, this area displays Shanghai's exploration and practice in three aspects, namely, greenness, low carbon and circulation.

Based on the dual-carbon goals, the multimedia on green, low-carbon circulation provides a large number of practical cases to explain carbon emissions peak and carbon neutrality in simple terms, showing a beautiful blueprint for Shanghai's green and low-carbon circulation in the future.

The exhibit on actions on environmental protection that have been rolled out in Shanghai since 2000 are highlighted. The interpretation of four major measures, namely, social co-government, harmonious coexistence, systematic pollution control, and green empowerment, demonstrates Shanghai's efforts in solving environmental problems and managing urban environment, pointing to the effectiveness of the actions.

The exhibit on garbage sorting presents the complete life garbage sorting system built under the principle of

Based on the indicators of ecological development, this area displays Shanghai's exploration and practice in three aspects, namely, greenness, low carbon and circulation.

Based on the dual-carbon goals, the multimedia on green, low-carbon circulation provides a large number of practical cases to explain carbon emissions peak and carbon neutrality in simple terms, showing a beautiful blueprint for Shanghai's green and low-carbon circulation in the future.

The exhibit on actions on environmental protection that have been rolled out in Shanghai since 2000 are highlighted. The interpretation of four major measures, namely, social co-government, harmonious coexistence, systematic pollution control, and green empowerment, demonstrates Shanghai's efforts in solving environmental problems and managing urban environment, pointing to the effectiveness of the actions.

The exhibit on garbage sorting presents the complete life garbage sorting system built under the principle of "reduction, resource, and harmlessness." The four vertical flow charts of the whole sorting process offer an intuitive view of the recycling of four types of garbage.

The exhibit on energy saving and low carbon focuses on green and low-carbon production and lifestyle. With data charts, it gives a clear picture of the utilization of shallow geothermal resources in building applications, outdoor heat exchange systems, ground source heat pump projects, and demonstrates the distribution of projects for developing and utilizing shallow geothermal energy in Shanghai.

展项解读

　　"绿色低碳循环"展项以 L 形展墙和多个互动科普展台营造出富有科技感的沉浸式体验空间，展项内设多种多媒体屏、模型及互动装置，将知识科普与体验互动的充分结合，为观众打造了一个智慧的科普展区。其中，碳中和游戏项目由 55 英寸显示屏、顶部条形 LED 屏、侧面显示屏及托盘交互装置组合构成，让观众在通关过程中提升节能低碳意识；由水上发电机组模型、建筑模型、导风管、手摇柄与风机配件等构成的风电装置丰富了观展互动性。

Interpretation

The L-shaped exhibition wall alongside several interactive stands for science popularization creates a technological immersive experience space. There are a variety of multimedia screens, replicas and interactive devices that combine knowledge with experience, offering the public a smart exhibition area on green and low-carbon circulation. The carbon neutrality game is available on a 55-inch display screen with a bar-shaped LED screen on the top, a side display screen and an interactive tray, so as to improve their awareness of energy saving and low carbon. The wind power device made up of replicas of water generating set, buildings, wind ducts, hand cranks and fan accessories, enriches the interactivity.

"绿色低碳循环"展项实景
"Green, Low-carbon Circulation" exhibit area

城市安全　Urban Security

" 城市安全 " 轴测图
Axonometric drawing of "Urban Security"

地面沉降防治查询屏 ———
Desk of land subsidence prevention and control.

展示内容

　　适应海平面上升，主动防治灾害是上海保障城市安全的紧迫任务，也是提升城市韧性、实现社会经济可持续发展的重要保证。"城市安全"从"城市防洪除涝""海绵城市建设""地面沉降防治"三方面展示上海建设韧性城市的主要举措。

Content

Responding to rising sea level and proactively preventing disasters are urgent tasks for Shanghai to ensure its urban security, as well as an important guarantee for improving urban resilience and achieving sustainable economic and social development. This area showcases Shanghai's main measures to build a resilient city from three aspects, namely, "urban flood and waterlogging control, " "construction of a sponge city" and "land subsidence prevention and control."

展项解读

　　"城市安全"展项结合两台触摸屏，生动演绎城市安全系列科普影片，为观众营造富有趣味性的观展空间。科普影片以 MG 及三维模拟动画等多样化数字形式，向观众形象展示上海城市安全防控治理情况。"地面沉降防治"触摸屏将上海地面沉降产生的危害、如何防治以及监测防治体系作出生动诠释。

Interpretation

Two touch screens vividly present videos about urban security in MG and 3D animation, creating an interesting exhibition space. The touch screen "Land Subsidence Prevention and Control" demonstrates the solution of Shanghai, its hazards and the monitoring and prevention system.

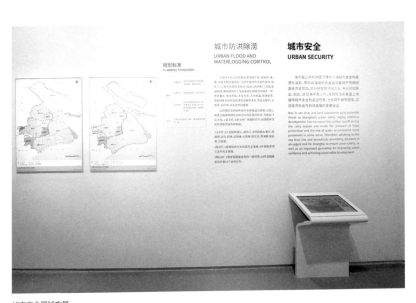

城市安全展项实景
"Urban Security" exhibit area

生态空间规划　Ecological Space Planning

"生态空间规划"轴测图
Axonometric drawing of "Ecological Space Planning"

自然保护地展台
Stand of nature reserve
construction

太浦河展台
Stand of Dapu River

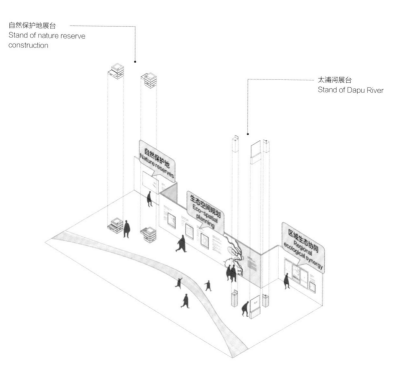

自然保护地
Nature reserves

生态空间规划
Eco-spatial
planning

区域生态协同
Regional
ecological synergy

展示内容

　　建设与具有世界影响力的社会主义现代化国际大都市相匹配的"城在园中、林廊环绕、蓝绿交织"的生态空间，打造一座令人向往的生态之城。"生态空间规划"展项重点展示"构建区域一体化生态格局""锚固市域生态网络结构""优化主城区蓝绿空间网络""加强刚柔并济的生态空间管控"四个方面的规划内容。

Content

Shanghai aims to build an ecological space with "city in the garden, forest corridor surrounded by trees, blue and green" so as to match the modern socialist metropolis with world influence, and to create a desirable ecological city. The "Ecological Space Planning" exhibition focuses on four aspects of planning: "Constructing an ecological framework of regional integration," "Consolidating citywide ecological network structure," "Optimize the blue-and-green space network of the main city," "Strengthen management and control of ecological space."

展项解读

　　"生态空间规划"展项位于展厅中央南侧区域，是四层展项的视觉中心，全幅面"生态空间规划图"结合凸字形展墙，突出上海生态空间的特色，延续了四层生态规划的脉络。展项整体呈线性平面设计，展墙内容随建筑墙体延伸，两侧展示自然保护地建设和区域生态协同的具体案例，并点缀布置相应展台。蓝色的地面"河流"与绿色的展台、展墙融合，形成一幅蓝绿交织的美好图景。

　　景展项通过大幅面雕版地图展示长江口地区东海海域环太平洋杭州湾等重点生态区域，在保证专业性和科学性的同时，增加了规划图纸的可阅读性。自然保护地展台采用"开盲盒"的游戏形式，盲盒内的设计结合自然保护地的影像图，不规则地植入相应生物物种的信息标签，让宏观而抽象的规划知识成为"唾手可得"的展品，使"盲盒"带来的惊喜之感点亮观众每一次的参观体验。

生态空间规划展墙设计示意图
The design of Ecological Space Planning

加强刚柔并济的生态空间管控
STRENGTHEN MANAGEMENT
AND CONTROL OF ECOLOGICAL SPACE

优化主城区蓝绿空间网络
OPTIMIZE THE BLUE-AND-GREEN SPACE
NETWORK OF THE MAIN CITY

锚固市域生态网络结构
CONSOLIDATING CITYWIDE ECOLOGICAL
NETWORK STRUCTURE

Interpretation

The "Ecological Space Planning" exhibit area is located in the central south area of the fourth floor as the visual center. The full-size "Ecological Space Planning Map" combined with the convex wall emphasizes the characteristics of Shanghai's ecological space and continues the lineage of ecological planning on the fourth floor. The exhibit is designed in a linear plan, with the content of the wall extending along with the building wall, showing concrete examples of nature reserve construction and regional ecological synergy on both sides, and decorated with corresponding display stands. The blue ground "river" blends with the green display stands and walls to form a beautiful picture of blue and green intertwined.

The exhibit showcases key ecological areas such as the Yangtze River estuary, the East China Sea, and the Pacific Rim of Hangzhou Bay through a large engraved map, which increases the readability of the planning drawings while ensuring professionalism and science. The design of the blind box is combined with the image map of the nature reserve, and the information labels of the corresponding biological species are irregularly implanted, so that the macro and abstract planning knowledge becomes an exhibit that can be readily accessed, which makes the "blind box" a very interesting experience. The "blind box" brings a sense of surprise that lights up the visitors' experience every time they visit.

城乡公园体系　Urban-rural Park System

"城乡公园体系"轴测图
Axonometric drawing of "Urban-rural Park System"

20

展示内容

　　结合市域空间布局，从服务能级角度，建立城乡一体、科学合理的公园分级体系。"城乡公园体系"展示了上海完善多层次城乡公园体系，大力推动环城生态公园带建设的系列举措。

　　结合市域空间布局，从服务能级角度，建立城乡一体、科学合理的公园分级体系。"城乡公园体系"展示了上海完善多层次城乡公园体系，大力推动环城生态公园带建设的系列举措。

　　展项全景式呈现了上海郊野公园总体布局和建设推进情况，重点介绍多功能农场型、远郊湿地型、郊野遗址文化型、近郊休闲型、远郊生态涵养型、滨江生态森林型、近郊都市森林型七类郊野公园规划理念、具体位置、风貌特征、资源特点、人文底蕴、产业配置、生态资源等信息。

Content

A scientific park grading system featuring integrated urban-rural development is established with the spatial layout of Shanghai and its services considered. The Urban-Rural Park System reveals the series of measures Shanghai adopts to improve the multi-level urban and rural park system and promote the ecological park belt around the city.

A panoramic view is given of the overall layout and construction progress of country parks in Shanghai, covering seven types, namely, multi-functional farm, outer-suburban wetland, country cultural heritage, suburban recreation, outer-suburban ecological conservation, riverside ecological forest, and suburban urban forest. Specifically, their planning, locations, style, resource characteristics, cultural heritage, industrial allocation, ecological resources and other information are provided.

城乡公园体系展项实景
"Urban-Rural Park System" exhibit area

郊野公园：广富林郊野公园
Country Park: Guangfulin Relics Park

城市公园：世纪公园
City Park: Century Park

地区公园：上海古猗园
District Park: Shanghai Guyi Garden

社区公园：桂林公园
Community Parks: Guilin Park

口袋公园：徐汇高安花园
Pocket Park: Xuhui Gao'an Park

多功能农场型：金山廊下郊野公园
Multi-functional Farm: Jinshan Langxia Country Park

远郊湿地型：青西郊野公园
Outer-suburban Wetland: Qingxi Country Park

郊野遗址文化型：广富林郊野公园
Country Cultural Heritage: Guangfulin Relics Park

近郊休闲型：嘉北郊野公园
Suburban Recreation: Jiabei Country Park

远郊生态涵养型：长兴岛郊野公园
Outer-suburban Ecological Conservation: Changxing Island Country Park

滨江生态森林型：松南郊野公园
Riverside Ecological Forest: Songnan Country Park

近郊都市森林型：浦江郊野公园
Suburban Urban Forest: Pujiang Country Park

展项解读

　　"城乡公园体系"展项充分利用展厅东南角的围合空间，结合三面展墙和两个 VR 互动座椅形成"城中有城，馆中有景，一草一木皆是情"的展览叙事空间。展墙采用"大型雕版地图 + 文字"的表达方式，辅以灰绿色的背景墙面，延续了"生态之城"的视觉设计元素。

Interpretation

The enclosed space in the southeast corner of the floor is combined with three exhibition walls and two VR interactive seats to create a narrative space with views. The exhibition walls feature "large engraved maps with texts" and a celadon background, which also feature the visual design elements of An Ecological City.

技术特色

　　展项采用 VR 设备和互动应用，以郊野公园的四季为主题，通过 360° 的 VR 技术全景装置，为观众打造云端沉浸式漫游体验。

　　观众戴上 VR 一体机便可身临其境地感受三维立体技术带来的视觉冲击——春风拂过金山廊下郊野公园的万亩良田和大地景观，夏季青西郊野公园里一片郁郁葱葱，秋季的嘉北郊野公园中油画一般的水乡美景，冬季的广富林郊野公园的文化韵味如同一幅水墨画。

Technical Features

With four seasons of country park as the theme, VR equipment and interactive applications are used to create an immersive cloud roaming experience for the audience.

The VR all-in-one machine allows the visitors to feel the visual impact brought by the 3DI technology—the spring breeze blowing over the huge fertile fields and earth in the Jinshan Langxia Country Park, the lush Qingxi Country Park in summer, and picturesque water scenery of Jiabei Country Park in autumn, and the cultural charm of Guangfulin Relics Park in winter.

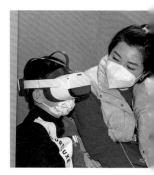
VR 一体机
VR experience available

乡村振兴
Rural Revitalization

"乡村振兴"轴测图
Axonometric drawing of "Rural Revitalization"

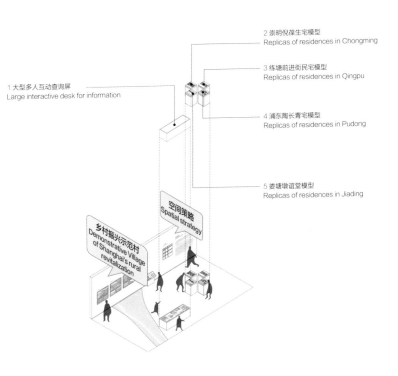

2 崇明倪葆生宅模型
Replicas of residences in Chongming

3 练塘前进街民宅模型
Replicas of residences in Qingpu

1 大型多人互动查询屏
Large interactive desk for information

4 浦东陶长青宅模型
Replicas of residences in Pudong

5 姜塘墩谊堂模型
Replicas of residences in Jiading

空间策略
Spatial strategy

乡村振兴示范村
Demonstrative Village
of Shanghai's rural
revitalization

展示内容

上海全面实施乡村振兴战略。"乡村振兴"展项立足中央决策部署和乡村振兴总体要求，重点展示了上海乡村振兴空间策略的制度创新，讲述了"冈身松江文化圈""淞北平江文化圈""沿海新兴文化圈"和"沙岛文化圈"四个文化圈层的特色，重点围绕风貌传承、产业振兴、生态宜居三种类型展示上海乡村振兴示范村案例和乡村建设的实践探索。

Content

The "Rural Revitalization" exhibit area let you take a glimpse into the central government's decision and overall requirements for rural revitalization, and focuses on the institutional innovation of Shanghai's rural revitalization spatial strategy, the characteristics of four cultural circles, namely, the "Gangshen Songjiang Cultural Circle," the "Songbei Pingjiang Cultural Circle," the "Coastal Emerging Cultural Circle" and the "Shadao Cultural Circle."

展项解读

"乡村振兴"展项由一面展墙、四个乡村住宅模型以及一个互动查询台构成，空间布局自然流畅。乡村住宅模型采用"切片"的叙述方式，选取上海典型的乡村建筑——崇明倪葆生宅、练塘前进街民宅、浦东陶长青宅、娄塘墩谊堂进行呈现。原木材质的模型肌理，结合 60 厘米见方的同材质模型底座，营造江南乡村建筑精巧、温暖、自然的空间氛围。考虑到模型的环视要求，四个模型与展墙之间保留了步行和观展空间，磨砂玻璃支撑的展台让模型与展墙内容联系得更加紧密，为多种观展角度留下可能性。

Interpretation

This area consists of an exhibition wall, four replicas of rural houses and an interactive inquiry desk, with a natural and smooth spatial layout. The replicas of rural houses adopt a "slice—and—dice" narrative, selecting typical rural buildings in Shanghai -- the Ni Baosheng House in Chongming, the Qianjin Street House in Liantang, the Tao Changqing House in Pudong, and the Dunyi Hall in Loutang -- for presentation. The replica texture of the original wood material, combined with the 60 square cm replica base of the same material, creates a delicate, warm and natural spatial atmosphere of the rural architecture in Jiangnan. Considering the requirement of circumnavigating the replicas, walking and viewing spaces are reserved between the four replicas and the exhibition wall, and the frosted glass-supported stand allows the replicas to be more closely connected with the contents of the exhibition wall, leaving the possibility for a variety of viewing angles.

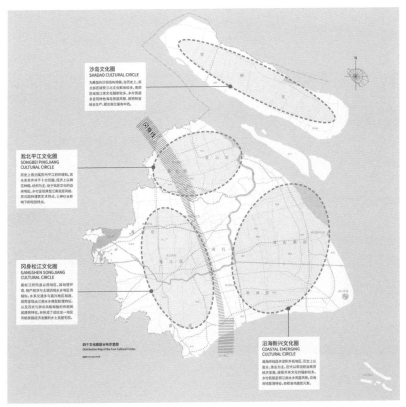

四个文化圈层分布示意图
Distribution map of the four cultural circles

浦东新区川沙陶宅
Tao's House on Chuansha of Pudong New Area

崇明堡镇倪葆生宅
Ni Baosheng's House on Chongming Island

青浦练塘前进街民宅
House on Qianjin Street, Liantang, Qingpu District

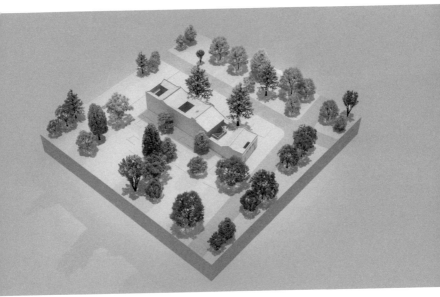

嘉定娄塘敦谊堂
Dunyitang Hall on Loutang of Jiading District

崇明世界级生态岛
Chongming World-class Ecological Island

"崇明世界级生态岛"轴测图
Axonometric drawing of "Chongming World-class Ecological Island"

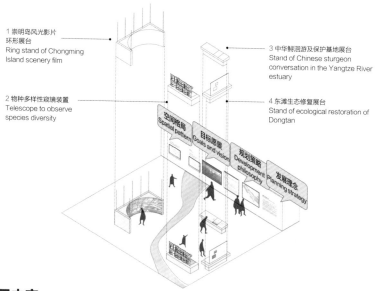

1 崇明岛风光影片
环形展台
Ring stand of Chongming
Island scenery film

2 物种多样性窥镜装置
Telescope to observe
species diversity

3 中华鲟洄游及保护基地展台
Stand of Chinese sturgeon
conversation in the Yangtze River
estuary

4 东滩生态修复展台
Stand of ecological restoration of
Dongtan

空间格局
Spatial pattern

目标愿景
Goals and vision

规划策略
Development
philosophy

发展理念
Planning strategy

展示内容

　　崇明岛是中国最大的河口冲击岛和沙岛，是 21 世纪上海可持续发展的重要战略空间。

　　"崇明世界级生态岛"展项围绕崇明建设具有全球引领示范作用的世界级生态岛的目标愿景，展示"+生态"和"生态 +"的规划策略、"三区两带两片"的空间格局、"五源多廊"的总体生态结构，通过展示生物多样性、长江口中华鲟保护、东滩生态修复等案例，反映崇明独特的生态底色、丰富的生态要素和规划实施的具体举措。

Content

Chongming Island is the largest estuarine impact island and sand island in China, and is an important strategic space for the sustainable development of Shanghai in the 21st century.

The "Chongming World-class Ecological Island" exhibit area focuses on Chongming's vision of building a world-class eco-island with global leadership and demonstration, and showcases the "+Eco" and "Eco+" planning strategy, the spatial pattern of "three zones, two belts and two areas," and the overall ecological structure of "five sources and multiple corridors," and reflects Chongming's unique ecological background, rich ecological elements and concrete measures of planning and implementation by displaying cases of biodiversity, conservation of Chinese sturgeon in the Yangtze River estuary and ecological restoration of Dongtan. The project will reflect Chongming's unique ecological context, diverse ecological elements and specific moves of implementation.

崇明世界级生态岛展项实景
"Chongming World-class Ecological Island" exhibit area

展项解读

  "崇明世界级生态岛"展项将知识科普作为展陈主线，采用展墙、互动投影、互动装置等多种互动途径，与文字、地图等多样呈现方式融为一体，同时结合《崇明之美》大幅面循环影片，徐徐展开一幅来自崇明的生态画卷。

  展项中央设有环形科普展示区，为观众提供沉浸式互动体验。展项四边各设有一个展台，分别以互动触摸屏、立式投影屏、悬挂式环形投影屏、魔方展台四种不同展陈形式呈现。

  互动触摸屏以中华鲟保护为案例，采用动画形式，利用投影互动装置，向观众形象展示有"水中大熊猫"之称的中华鲟的生活环境和洄游路线。观众可以通过投影画面动态了解长江口中华鲟保护的意义和举措，通过触摸集成纳米触控显示屏查询具体信息。

  立式投影屏由两台投影仪、两部红外摄像头、两个吸顶扬声器构成完整的互动投影系统，动态展示了崇明岛候鸟的迁徙路线以及东滩修复的过程。

  悬挂式环形投影屏在视觉上营造出观展的空间包裹感，带领观众走进崇明东滩鸟类国家级自然保护区，拥抱自然。

  魔方展台利用三台望远镜为观众创造出近距离观察鸟类迁徙觅食的真实场景。生物多样性的磨方设计融入了"阅读—解谜"的游戏体验，观众拨动旋转展架上悬挂的28个立方体，鸟类剪影、介绍、栖息地环境等信息逐次展开，抽丝剥茧般满足观众的认知好奇心。

Interpretation

This area takes knowledge and science as the main line of exhibition, and adopts various interactive ways such as exhibition wall, interactive projection and interactive device, and integrates with various presentation methods such as text and map, and at the same time combines with "The Beauty of Chongming" large format loop film to slowly unfold an ecological picture of Chongming.

At the center of the space is a circular science education area, which provides an immersive and interactive experience. There is a display stand on each of the four sides, each assuming a screen in a different form: interactive touch, vertical projection, hanging ring projection and magic cube display stand.
Citing the protection of Chinese sturgeon as an example, the interactive touch screen uses animation to show visitors the living environment and migratory route of the sturgeon, which is known as the "giant panda in the water." Visitors can learn about the significance and initiatives of Chinese sturgeon conservation in the Yangtze River Estuary through the projection screen, and inquire about specific information by touching the integrated nano-touch display.

The vertical projection system, which consists of pairs of projectors, infrared cameras and ceiling speakers, animatively displays the migration route of migratory birds on Chongming Island and Dongtan's restoration process.

The suspended ring projection screen creates a sense of spatial wrapping for viewing; affording you a virtual experience of the Chongming Dongtan Bird National Nature Reserve.

The Rubik's Cube display stand uses three telescopes to create a realistic scenario for visitors to observe birds migrating and feeding up close. The design of the biodiversity mill cube incorporates a "read and crack" game experience, in which visitors poke the 28 cubes hanging on the rotating display, and information such as bird silhouettes, introductions, and habitat environments unfold one by one, satisfy your cognitive curiosity.

魔方展台
View of the Magic Cube display stand

观众通过不同高度的望远镜来观察崇明东滩的自然世界
Telescopes of different height to watch the world of the Chongming Dongtan Bird National Nature Reserve

技术特色

　　立式投影屏展台采用透明屏的投影展示技术，运用一块长 2.37 米、宽 1.4 米的通电玻璃来作为投影介质。通电玻璃是由两块玻璃中间夹胶一整块通电膜而构成，利用其特点可在展示过程中增加变化层次，巧妙融合鸟类迁徙路线与东滩生态修复图景的同时，确保观展动线对空间通透感的要求。

Technical Features

The stand with vertical projection adopts the projection technology of transparent screen, using a 2.37-m by 1.4-m piece of energized glass as the medium. The energized glass, composed of an energized film laminated between two glass panels, can add layers of changes in the display process, cleverly integrating the bird migration route and the ecological restoration scenery of Dongtan while ensuring that the viewing line's space permeability.

"上海2035"：规划实施框架体系
"Shanghai 2035" :
The Framework System of Implementation

"'上海2035'：规划实施框架体系"轴测图
Axonometric drawing of "'Shanghai 2035' :The Framework System of Implementation"

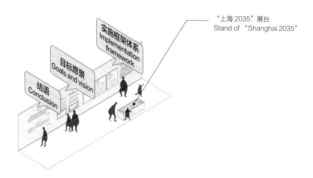

"上海2035"展台
Stand of "Shanghai 2035"

"'上海2035'：规划实施框架体系"展项设计图
Graphic of " 'Shanghai 2035' The Framework System of Implementation" exhibit area

展示内容

习近平总书记指出："展望未来，我们完全有理由相信，在新时代中国发展的壮阔征程上，上海一定能创造出令世界刮目相看的新奇迹，一定能展现出建设社会主义现代化国家的新气象！"

上海市坚持一张蓝图干到底，以钉钉子精神抓好规划的组织实施。"'上海2035'：规划实施框架体系"展项展示了上海建立空间维度、时间维度、政策维度的规划实施框架体系，反映上海坚持以人民为中心，走出符合超大城市特点的内涵式、集约型、绿色化的"上海2035"高质量实施之路。

Content

"Looking ahead, we have every reason to believe that in the great journey of China's development in the new era, Shanghai will create new wonders that will impress the world and turn a new chapter in building a modern socialist country," says President Xi Jinping.

展望 | 2035年
2035, A YEAR OF ENVISIONING

基本建成卓越的全球城市，令人向往的创新之城、人文之城、生态之城，具有世界影响力的社会主义现代化国际大都市。描摹并彰显标志性指标的先水平，在我国基本实现社会主义现代化的进程中，始终处好新时代发展的排头兵、改革发展的先行者。

Shanghai will basically build itself into an excellent global city, an admirable city of innovation, humanity and sustainability as well as a modern socialist international metropolis with world influence. With the critical indicators of development reaching world-leading level, the city will serve as a vanguard and pioneer in reform and opening-up and innovation throughout the process of basically building a modern socialist country.

梦圆 | 2050年
2050, A YEAR OF ACCOMPLISHING

全面建成卓越的全球城市，令人向往的创新之城、人文之城、生态之城，具有世界影响力的社会主义现代化国际大都市。各项发展指标全面达到国际领先水平，为我国建成富强民主文明和谐美丽的社会主义现代化强国，实现中华民族伟大复兴中国梦谱写更新的壮丽上海篇章。

Shanghai will complete the building of an excellent global city, an admirable city of innovation, humanity and sustainability in all respects as well as a modern socialist international metropolis with world influence. With all indicators of development reaching world-leading level, the city will write a more brilliant chapter for building China into a great modern socialist country that is prosperous, strong, democratic, culturally advanced, harmonious, and beautiful and realizing the Chinese Dream of national rejuvenation.

结 语

展望未来，我们完全有理由相信，在新时代中国发展的壮阔征程上，上海一定能创造出令世界刮目相看的新奇迹，一定能展现出建设社会主义现代化国家的新气象！
——习近平

CONCLUSION

Looking ahead, we have every reason to believe that in the great journey of China's development in the new era, Shanghai will create new wonders that will impress the world and turn a new chapter in building a modern socialist country!
——XI JINPING

"Shanghai adheres to a blueprint to the end, and grasps the organization and implementation of the plan with the spirit of nailing down." The exhibit area " 'Shanghai 2035' :The Framework System of Implementation" shows the establishment of a spatial, temporal and policy-oriented planning and implementation framework system in Shanghai, reflecting Shanghai's continued adherence to a people-centered, connotative, intensive and green implementation path that is in line with the characteristics of a mega-city. The exhibition reflects Shanghai's continued adherence to the people-centered approach and the high quality implementation of "Shanghai 2035" in a connotative, intensive and green way that meets the characteristics of a mega-city.

展项解读

从首层序厅的"上海印象"开始，到二至四层讲述人文之城、创新之城、生态之城，最终落脚于"'上海2035'：规划实施框架体系"，从观展流线和内容设计的角度对全馆展陈进行了首尾呼应。展项由展墙和查询台组成。多人互动查询台记录了"上海2035"规划和上海人民意见征集的相关内容，观众在屏幕内留下观展体验，也留下了对这座城市的祝福与希冀。

Interpretation

Starting from the "Shanghai Impression" in the preface hall, to the "Humanistic City," the "Innovative City," and the "Ecological City" on floors two through four, and finally ends with "Shanghai 2035," the tour is well designed in terms of the route and the holdings. Each exhibit area consists of an exhibition wall and an information desk. The interactive information desk collects visitors' opinions about the "Shanghai 2035" plan and the museum, and visitors leave their viewing experience on the screen, as well as their wishes and hopes for the city.

特殊展项

Special Exhibits

5D 数字化沉浸式城市沙盘
5D Digital Immersive
City Sandtable

展项轴测图
Axonometric drawing

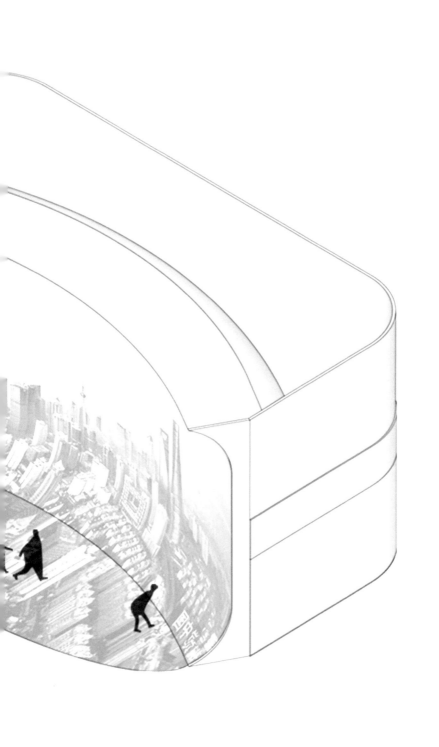

展项解读

　　"5D数字化沉浸式城市沙盘"运用数字孪生理念，构建可交互的大型沉浸式混合现实场景，高度融合数字化城市展示与分析模拟功能，是城市规划展示的"智慧大脑"。

　　沙盘以数字化手段综合反映城市规划发展的目标与愿景，从内容上包括城市发展主题演绎和城市发展专题演绎两类，从服务对象上包括公众、VIP和专家三类群体，核心是为公众提供更全面、更清晰的展示服务。

Interpretation

The "5D Digital Immersive City Sandtable" uses the concept of digital twin to build a large interactive immersive mixed reality scene, which is highly integrated with digital city display and analysis simulation functions, and is the "intelligent brain" of urban planning display.

The Sandtable reflects the goal and vision of urban planning and development by digital means, including two types of urban development theme interpretation and urban development features interpretation from the content, and three types of groups including public, VIP and experts from the service object, with the core of providing more comprehensive and clearer display services for the public.

展项解读

"5D 数字化沉浸式城市沙盘"位于展厅中央，纵贯三、四两个楼层，巨大的内凹双曲面投影屏幕构建出大型沉浸式展示场景，生动、真实地动态演绎了城市空间。相比传统实体沙盘，数字沙盘尝试突破空间和时间的限制，力图通过数字技术展示更全面的城市发展图景。

Interpretation

Running through the third and fourth floors, the "5D Digital Immersive City Sandtable" is a must-stop at the center of the galley, where a huge concave hyperbolic projection screen a large immersive display scene, vividly and realistically interpreting the urban space dynamically. Compared with a traditional sandtable, the digital sandtable pushes the limits of space and time and gives a panoramic view of the city by using digital technology.

"5D 数字化沉浸式城市沙盘"实景
5D Digital Immersive Urban Sandtable

城市实验室 Urban Lab

　　"城市实验室"定位为"教育·体验·探索"，突出"交互体验"，是全国首家基于真实地理信息数据的空间规划实验平台，是以专业研究和科普教育为目的共享空间，通过实验方式引导公众探索城市发展的各种可能。

　　城市实验室主要面向专业人士与青少年群体。为专业人士提供研讨、专业设计和成果交流的平台；为青少年打造城市规划系列科普互动课程、开展创新设计竞赛等活动。

"城市实验室"轴测图
Axonometric drawing of the "Urban Lab"

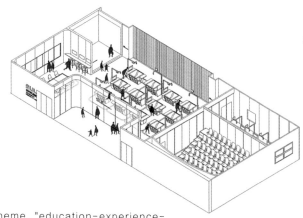

Embracing the theme, "education-experience-exploration," the Urban Lab is a space for professional research and science education alike. It affords intriguing interactive experience, and encourages the visitors to explore various possibilities of urban development through experiments. It is the first experiment platform of its kind in China that is equipped with real geographic data.

The Urban Lab is a go-to place for both professionals and teenagers alike: It allows the former to discuss, design and brainstorm, while its interactive urban planning courses and innovative design competitions beckon to the latter.

"城市实验室"实景
Exhibit area of the "Urban Lab"

展项解读

"城市实验室"位于二楼展厅东侧，占地约400平方米，由信息查询区、分享交流区、实验操作区和成果展示区四部分组成。

信息查询区主要展示全球主要城市的发展战略、各类主题的城市规划研究成果，以及政府已公开的重要规划信息。该区域设有墙面和桌面多媒体查询设备，技术上实现多点触控，同时定制案例互动程序，有效支持图片的缩放和移动。

Interpretation

The Urban Lab is located on the east side of the second floor, covering an area of some 400 square meters. It consists of four areas, carved for the purposes of information inquiry, exchange, experiment & operation and results display.

The Information Area is your cup of tea if you want to consult development strategies of major cities around the world, urban planning research results on various themes, and important planning intelligence disclosed by governments. Equipped with wall and desktop multimedia inquiry devices, the area can technically realize multi-touch control; its case-specific interactive program allows you to zoom in and move the pictures.

实验操作区是专为青少年打造的数字化规划实验区，青少年通过"角色扮演"的方式参与了解城市规划项目的全过程。定制开发的互动实验桌通过智能识别技术，并结合真实的地理信息数据为参与者营造沉浸式的实验空间。每张实验桌都配有感应识别系统，通过快速扫描识别模拟沙盘并智能分析，实时反馈规划设计成果和规划指标数据，直观明了。

成果展示区是由四面 LED 屏围合而成 10 平方米的"CAVE 空间"，体验区，实验操作区完成的成果通过三维多通道场景还原与展示技术可直接在该区域呈现，参与者可在三维空间里直观感受实验成果的实时渲染效果。

The Experiment & Operation Area is a digital planning experiment area tailored for young people; they can attend the whole process of urban planning projects through "role-playing." The custom-made interactive experiment tables create an immersive experiment space for participants through intelligent recognition technology combined with real geographic information data. Each table is equipped with an inductive recognition system, which provides real-time feedback on planning and design results and planning index data through rapid scanning and recognition of the simulation Sandtable and intelligent analysis, with intuitive and clear results.

The Results Display Area is a 10-square-meter "CAVE space" surrounded by four LED screens. The results generated in the experimental operation area completed through the 3D multi-channel scene restoration and display technology can be presented directly in the area, participants can intuitively feel the real-time rendering effect of experimental results.

实验操作区
The Experiment & Operation Area

地下空间
Temporary Exhibition Hall (B1)

地下一层主要包括临展厅和城市通道。

临展厅采用弹性的空间设计，展示空间可根据需求灵活布局，未来将不定期举办规划公示、城市设计、文化艺术等专题类展览，观众可在此了解最新城市规划研究成果，欣赏到国内外艺术作品。

城市通道以江水为主题，以白色和蓝色为主色调。通道内的吊顶和墙壁采用参数化设计的波浪 GRC 造型模块塑造出波澜壮阔的浪潮，象征着革新浪潮涌动下，上海不断奋发前行的创新动力和奋斗精神。

城市通道如同微缩的水岸之城
The "city passageway" is like a miniature share-based city

城市通道中的波浪造型
Wave-shaped wall and device

城市通道内涌动的"浪花"
Waves and waves

The basement houses the temporary exhibition hall and the "city passageway."

The temporary exhibition hall adopts flexible space design, and the display space can be flexibly laid out according to the demand. In the future, it will accommodate exhibitions of planning public announcements, urban design, culture and art, etc. The public can enjoy domestic and foreign art works and learn about the updates on urban planning research.

The river-themed "city passageway" features white and blue as the main colors. The ceiling and walls are shaped by parametrically designed wave GRC modeling modules, symbolizing the surging waves in Shanghai's waterways and the steady progress that the city craves for.

服务信息　Service Information

开放时间

9:00-17:00（16:00 停止入馆）
每周三闭馆（法定节假日除外）

Opening Hours

9:00-17: 00 (Last entry at 16:00)
Closed on Wednesdays (Excluding statutory holidays)

参观服务　Visit

预约方式 Reservation

　　观众可关注上海城市规划展示馆官方微信号了解
更多服务信息。

Please follow our official WeChat account for more
information.

官方微信公众号二维码
SUPEC Wechat

参观线路推荐 Recommended Routes

参观时长：90 分钟   Duration: 90 min

1F 序厅（上海印象、上海历次城市总体规划概述）（10分钟）

2F 人文之城（城市肌理、风貌格局、城市更新、黄浦江、苏州河：迈向具有全球影响力的世界级滨水区、城市设计标准、15 分钟社区生活圈、公共服务、公众参与）（30 分钟）

3F 创新之城（数字沙盘、自贸试验区、长三角一体化发展、五个新城、科创中心、服务提升、国际枢纽、城市数字化转型）（30 分钟）

4F 生态之城（岁月的积淀、广域的空间、底线约束、绿色低碳循环、城市安全、生态空间规划、城乡公园体系、乡村振兴、崇明世界级生态岛、"上海 2035"规划实施框架体系）（20 分钟）

直达电梯 B1 层离馆

1F Preface (Shanghai Impression and Shanghai Master Plans) (10 min)

2F A Humanistic City (Urban Fabric, Feature Patterns, Urban Regeneration, The River and the Creek, Urban Design Standards, 15-minute Community Life Circle, Public Services and Public Participation) (30 min)

3F An Innovative City (The Digital Sandtable, China (Shanghai) Pilot Free Trade Zone, Integrated Development of the Yangtze River Delta, Five New Cities, Science and Technology Innovation Center, Service Improvement, International Hub, and Urban Digital Transformation) (30 min)

4F An Ecological City (Precipitation of Ages, Extensive Space, Bottom-line Control, Green and Low-carbon Circulation, Urban Security, Ecological Space Planning, Urban-rural Park System, Rural Revitalization, Chongming World-class Ecological Island, "Shanghai 2035": The Framework System of Implementation) (20 min)

Take the elevator to B1 for exit

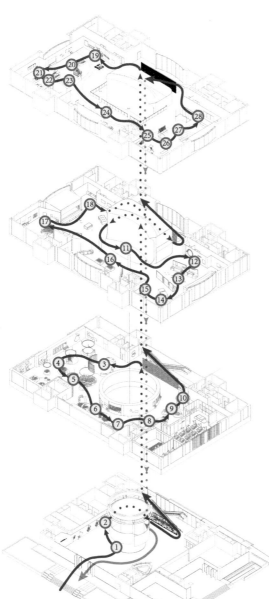

参观时长：90分钟
Duration: 90 min

1. 上海印象
Shanghai Impression

2. 上海历次城市总体规划概述
Shanghai Master Plans

3. 城市肌理
Urban Fabric

4. 风貌格局
Feature Patterns

5. 城市更新
Urban Regeneration

6. 黄浦江、苏州河：迈向具有全球影响力的世界级滨水区
The River and the Creek

7. 城市设计标准
Urban Design Standards

8. 15分钟社区生活圈
15-minute Community Life Circle

9. 公共服务
Public Services

10. 公众参与
Public Participation

11. 数字沙盘
The Digital Sand-table

12. 自贸试验区
China (Shanghai) Pilot Free Trade Zone

13. 长三角一体化发展
Integrated Development of the Yangtze River Delta

14. 五个新城
Five New Cities

15. 科创中心
Science and Technology Innovation Center

16. 服务提升
Service Improvement

17. 国际枢纽
International Hub

18. 城市数字化转型
Urban Digital Transformation

19. 岁月的积淀
Precipitation of Ages

20. 广域的空间
Extensive Space

21. 底线约束
Bottom-line Control

22. 绿色低碳循环
Green and Low-carbon Circulation

23. 城市安全
Urban Security

24. 生态空间规划
Ecological Space Planning

25. 城乡公园体系
Urban-rural Park System

26. 乡村振兴
Rural Revitalization

27. 崇明世界级生态岛
Chongming World-class Ecological Island

28. "上海2035"规划实施框架体系
"Shanghai 2035"-The Framework System of Implementation

直达电梯B1楼离馆
Take the elevator to B1 for exit

参观时间：60 分钟　Duration: 60 min

1F 序厅（上海印象、上海历次城市总体规划概述）（10 分钟）

2F 人文之城（城市肌理、风貌格局、黄浦江、苏州河：迈向具有全球影响力的世界级滨水区、15 分钟社区生活圈）（20 分钟）

3F 创新之城（自贸试验区、长三角一体化发展、五个新城、国际枢纽、数字沙盘）（20 分钟）

4F 生态之城（岁月的积淀、广域的空间、生态空间规划、崇明世界级生态岛）（10 分钟）

直达电梯 B1 层离馆

1F Preface (Shanghai Impression, and Shanghai Master Plans) (10 min)

2F A Humanistic City (Urban Fabric, Feature Patterns, The River and the Creek, 15-minute Community Life Circle) (20 min)

3F An Innovative City (China (Shanghai) Pilot Free Trade Zone, Integrated Development of the Yangtze River Delta, Five New Cities, International Hub and Urban Digital Transformation and The Digital Sandtable) (20 min)

4F An Ecological City (Precipitation of Ages, Extensive Space, Ecological Space Planning, Chongming World-class Ecological Island) (10 min)

Take the elevator to B1 for exit

239

参观时长：60分钟
Duration: 60 min

1.上海印象
Shanghai Impression

2.上海历次城市总体规划概述
Shanghai Master Plans

3.城市肌理
Urban Fabric

4.风貌格局
Feature Patterns

5.黄浦江、苏州河：迈向具有全球影响力的世界级滨水区
The River and the Creek

6.15分钟社区生活圈
15-minute Community Life Circle

12.自贸试验区
China (Shanghai) Pilot Free Trade Zone

13.长三角一体化发展
Integrated Development of the Yangtze River Delta

14.五个新城
Five New Cities

17.国际枢纽
International Hub

18.城市沙盘
The Digital Sand-table

19.岁月的积淀
Precipitation of Ages

20.广袤的空间
Extensive Space

24.生态空间规划
Ecological Space Planning

27.崇明世界级生态岛
Chongming World-class Ecological Island

直达电梯B1楼离馆
Take the elevator to B1 for exit

参观时间：45 分钟　Duration: 45 min

1F 序厅（上海印象、上海历次城市总体规划概述）（10分钟）

2F 人文之城（风貌格局、黄浦江、苏州河：迈向具有全球影响力的世界级滨水区）（10分钟）

3F 创新之城（长三角一体化发展、五个新城、数字沙盘）（15分钟）

4F 生态之城（岁月的积淀、广域的空间、生态空间规划、崇明世界级生态岛）（10分钟）

直达电梯 B1 层离馆

1F Preface (Shanghai Impression, and Shanghai Master Plans) (10 min)

2F A Humanistic City (Feature Patterns, and The River and the Creek) (10 min)

3F An Innovative City (Integrated Development of the Yangtze River Delta, Five New Cities, and The Digital Sandtable) (15 min)

4F An Ecological City (Precipitation of Ages, Extensive Space, Ecological Space Planning, Chongming World-class Ecological Island) (10 min)

Take the elevator to B1 for exit

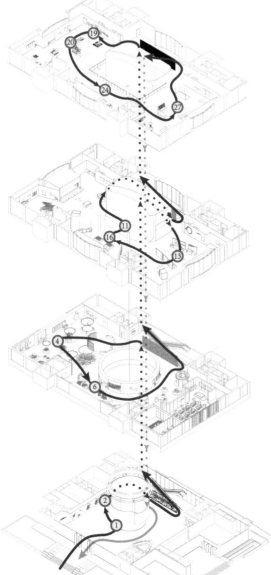

参观时长：45分钟
Duration: 45 min

1.上海印象
Shanghai Impression

2.上海历次城市总体规划概述
Shanghai Master Plans

4. 风貌格局
Feature Patterns

6. 黄浦江、苏州河：迈向具有全球影响力的世界级滨水区
The River and the Creek

13. 长三角一体化发展
Integrated Development of the Yangtze River Delta

14. 五个新城
Five New Cities

18. 数字沙盘
The Digital Sand-table

19. 岁月的积淀
Precipitation of Ages

20.广域的空间
Extensive Space

24.生态空间规划
Ecological Space Planning

27.崇明世界级生态岛
Chongming World-class Ecological Island

直达电梯B1楼离馆
Take the elevator to B1 for exit

公共服务设施　Public Service Facilities

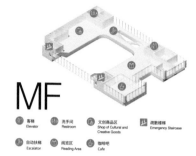

MF

客梯 Elevator	洗手间 Restroom	文创商品区 Shop of Cultural and Creative Goods	疏散楼梯 Emergency Staircase
自动扶梯 Escalator	阅览区 Reading Area	咖啡吧 Cafe	

1F

客梯 Elevator	咨询服务 Consultation Service	售票处 Ticket Office	入口 Entry
自动扶梯 Escalator	医务室 Medical Service	自助售票 Automatic Ticket	出口 Exit
疏散楼梯 Emergency Staircase			

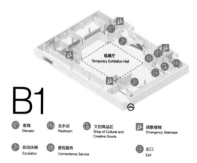

临展厅
Temporary Exhibition Hall

B1

客梯 Elevator	洗手间 Restroom	文创商品区 Shop of Cultural and Creative Goods	疏散楼梯 Emergency Staircase
自动扶梯 Escalator	便民服务 Convenience Service	出口 Exit	

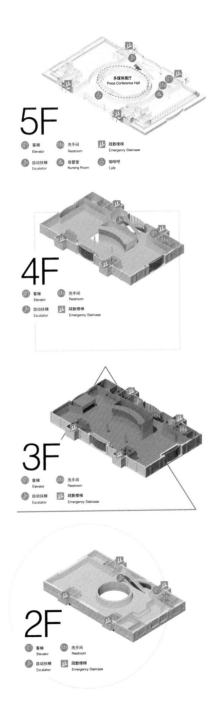

5F

客梯 Elevator 　洗手间 Restroom 　疏散楼梯 Emergency Staircase

自动扶梯 Escalator 　母婴室 Nursing Room 　咖啡吧 Cafe

4F

客梯 Elevator 　洗手间 Restroom

自动扶梯 Escalator 　疏散楼梯 Emergency Staircase

3F

客梯 Elevator 　洗手间 Restroom

自动扶梯 Escalator 　疏散楼梯 Emergency Staircase

2F

客梯 Elevator 　洗手间 Restroom

自动扶梯 Escalator 　疏散楼梯 Emergency Staircase

1F 服务中心   1F Service Center
咨询服务、导览服务、志愿者服务、医疗联络、失物招领、广播服务、意见处理、便民服务。
Consulting, guides, volunteers, medical service, lost and found, broadcasting, opinions center, and other services.

BF 服务中心   B1 Service Center
咨询服务、导览器归还服务、意见处理、寄存服务、便民服务。
Consulting, audio guides, opinions center, left luggage, and other services.

MF 文创及阅览区
Cultural Products & Reading Area
展示上海特色文创产品、提供最新城市规划与城市文化书籍供观众查阅。
Shanghai's unique cultural and creative products and the latest publications on urban planning and culture are available.

地址和交通
Address and Transportation

上海城市规划展示馆位于上海市人民大道 100 号（近西藏中路），毗邻南京路，与上海大剧院、上海博物馆等相邻。

The SUPEC is located at 100 Renmin Avenue, Shanghai (near Middle Xizang Road), adjacent to Nanjing Road, Shanghai Grand Theater and Shanghai Museum.

地铁 1、2、8 号线 2 号出口，公交 49 路、18 路等人民广场站。
Gate 2 of Shanghai Metro Lines 1, 2 and 8. People's Square stop of No. 49 and No. 18 buses.

插图	刘培爽			Picture	Liu Peishuang		
绘图	朱辰辰	郭小溪		Drawing	Zhu Chenchen	Guo Xiaoxi	
摄影	郑宪章	徐捷	刘杰	Photograph	Zheng Xianzhang	Xu Jie	Liu Jie
文字	苏杭	郭小溪		Text	Su Hang	Guo Xiaoxi	
翻译	郑超			Translator	Zheng Chao		

图书在版编目（CIP）数据

上海城市规划展示馆参观指南：汉英对照 / 上海城
市规划展示馆编. -- 上海：上海文化出版社，2022.12
　ISBN 978-7-5535-2658-4

　Ⅰ.①上… Ⅱ.①上… Ⅲ.①城市规划－展览馆－上
海－指南－汉、英 Ⅳ.①TU984.251-28

中国版本图书馆CIP数据核字(2022)第233935号

出　版　人：姜逸青
责任编辑：江　岱　张　彦
装帧设计：王　伟

书　　名：**上海城市规划展示馆参观指南**（汉英对照）
　　　　　上海城市规划展示馆　编
出　　版：上海世纪出版集团 上海文化出版社
地　　址：上海市闵行区号景路159弄A座三楼　201101
发　　行：上海文艺出版社发行中心
　　　　　上海市闵行区号景路159弄A座二楼　201101 www.ewen.co
印　　刷：上海雅昌艺术印刷有限公司
开　　本：889×1194　1/32
印　　张：7.75
印　　次：2022年12月第1版 2022年12月第1次印刷
书　　号：ISBN 978-7-5535-2658-4/TU.016
定　　价：128.00元
告 读 者：如发现本书有质量问题请与印刷厂质量科联系 T：021-68798999